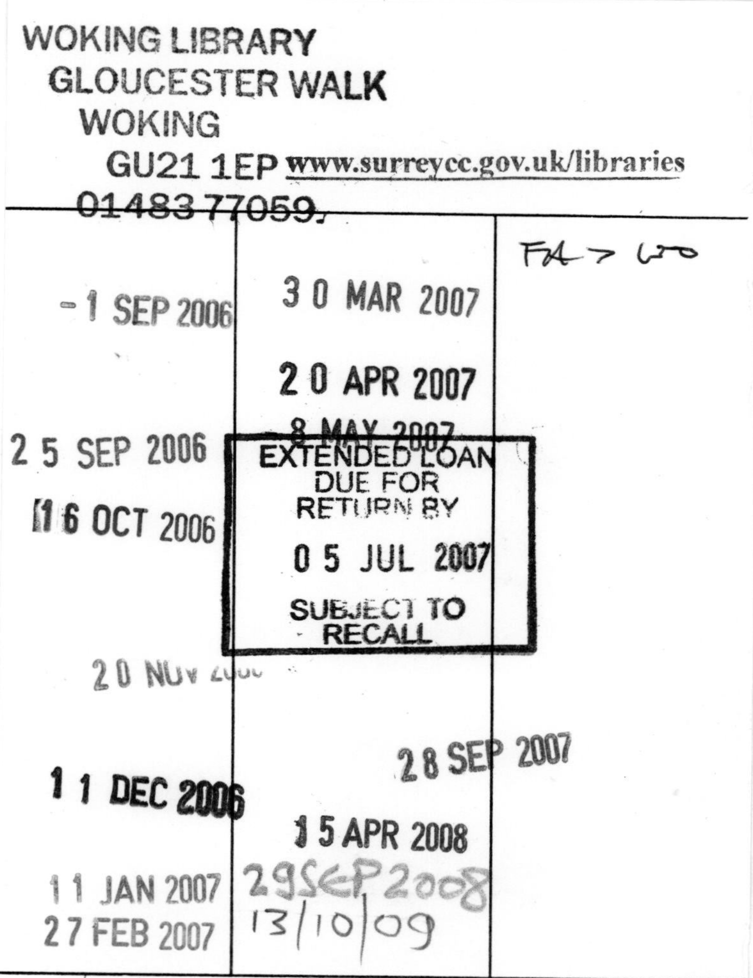

MESSERSCHMITT Me 264

AMERIKA BOMBER

The Luftwaffe's Lost Transatlantic Bomber

MESSERSCHMITT Me 264

AMERIKA BOMBER

The Luftwaffe's Lost Transatlantic Bomber

ROBERT FORSYTH
and Eddie J. Creek

An imprint of
Ian Allan Publishing

Messerschmitt Me 264 *Amerika Bomber* is the result of a collaboration between Robert Forsyth and Eddie Creek, both of whom have collected documentary and photographic material on the aircraft for several years. Eventually, they decided to do something with what they had found.

Robert Forsyth is the author of *JV 44 – The Galland Circus* (1996), *Battle over Bavaria: The B-26 versus the German Jets* (1998) and *Mistel: German Composite Aircraft and Operations 1942-1945* (2001).

Eddie J. Creek is co-author of a number of landmark books on Luftwaffe aircraft including *Jet Planes of the Third Reich* (1982), *Arado Ar 234* (1992), *Me 262* (1997-2000 - four volumes) and *On Special Missions – The Luftwaffe's Research and Experimental Squadrons 1923-1945* (2003). He has contributed to more books than he cares to remember.

In 1995 the authors set up the Classic Publications imprint, which is now owned by Ian Allan Publishing. They now run an independent book production company and have produced more than 70 acclaimed aviation titles.

First published 2006

ISBN (10) 1 903223 65 2

ISBN (13) 978 1 903223 65 9

Produced by Chevron Publishing Limited

Cover and book design by Colin Woodman Design

An imprint of Ian Allan Publishing Ltd, Hersham, Surrey KT12 4RG.

Printed in England by Ian Allan Printing Ltd, Hersham, Surrey KT12 4RG.

Visit Ian Allan Publishing at www.ianallanpublishing.com

Contents

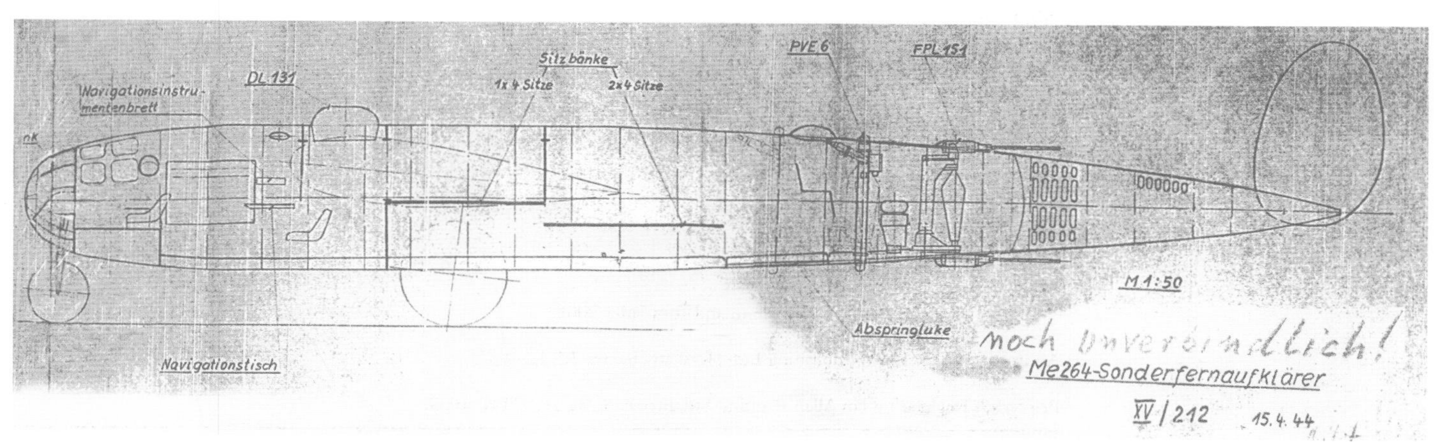

Navigationsinstrumentenbrett
DL 131
Sitzbänke
1x4 Sitze
2x4 Sitze
PVE 6
FPL 151
M 1:50
Absprungluke
Navigationstisch
noch unverbindlich!
Me264-Sonderfernaufklärer
XV/212
15.4.44

Introduction

The notion of Nazi Germany possessing a strategic, long-range bomber capable of delivering a significant offensive load against the eastern seaboard of the United States or Moscow and other key areas of American and Soviet industrial output, has long been a source of absorbing speculation to the aviation historian with cause for considerable conjecture.

Some time during the early years of the Second World War, the term '*Amerika Bomber*' was given to the Messerschmitt Me 264. Although the aircraft's designer, *Professor* Willy Messerschmitt, secured his place in aviation history for his famous Bf 109 fighter and the Me 262, the world's first jet fighter, the Me 264, though relatively unknown, remained very much a personal quest for Messerschmitt – the challenge of range over weight; the challenge to design and build an aircraft that would possess sufficient endurance to reach the Antipodes from Germany or to fly from any point of the globe to another without refuelling.

The onset of war in 1939, and the threat to Germany from the United States' entry into that conflict two years later, meant that Messerschmitt had the opportunity to revisit and exploit his design as a long-range, *transatlantic* bomber and reconnaissance aircraft. The concept was no less revolutionary than the Me 262.

As the war progressed, '*Amerika Bomber*' was also adopted as a more generic term for designs from other aircraft manufacturers which had the capability – on paper at least – of being able to reach North America from Europe and return. It was the aircraft for which Hitler and senior German air commanders hungered.

However both during the war and after it, the term developed into something almost mythical, attracting exaggeration, inaccuracy and falsification.

We do not attempt to portray this study as 'the last word'. Rather, the purpose of this book is to present the reader with an insight as accurate as possible into the intriguing, complex, contradictory, and at times downright farcical story of the Me 264 – the *Luftwaffe's* 'lost' transatlantic bomber – based on the information we have so far located.

During the course of our research, we have been fortunate enough to be able to draw upon many wartime documents relative to the Me 264's development such as flight test reports and minutes of key meetings, as well as associated drawings and many fascinating photographs. We have endeavoured to include as much as we can of the latter in the book you now hold.

Little would have been possible without the kind help and generosity of a number of individuals to whom we would like to express our sincere thanks.

Firstly, however, we must acknowledge the work of Manfred Griehl whose own research on the Me 264 during the 1990s has proven immensely valuable to our own work. We recommend Manfred's articles and books as an excellent source of further reading and details of these can be found in the sources and bibliography at the back of this book.

In the UK we would like to thank in particular Mr Stephen Walton of the Department of Documents at the Imperial War Museum for allowing us access to hundreds of feet of microfilm which revealed much new information.

We are also grateful to the following for their kind assistance with this project: Giorgio Apostolo and AeroFan Journal, Isolde Baur, Karl Bleckmann, Steve Coates, John Fedorowicz, Giancarlo Garello, David Irving, Antony L. Kay, Gino Marcomini, Mike Norton, Mike Olive, Huib Ottens, Dr Jochen Prien, Dr Alfred Price, Günter Sengfelder, J. Richard Smith, Erwin Schnetzer, Tom Tullis, Colin Woodman and EADS Corporate Heritage.

We hope that this book represents just the beginning of a deeper understanding of this fascinating aircraft.

Robert Forsyth
Eddie J Creek
March 2006

Glossary

Abteilung	Detachment or section
ACTS	Air Corps Tactical School (US)
Aerodynamische Versuchsanstalt (AVA)	Aerodynamic Test Establishment
Akaflieg	Academic flying club or association
Alpenflug	Alpine flying race
Amtsgruppe Entwicklung	Development Office (RLM)
Baubeschreibung	Construction description
Befehlshaber der U-boote (BdU)	Commander U-boats
Blindflugschule	Blind Flying School
Deutsche Akademie der Luftfahrtforschung	German Academy for Aviation Research
Deutsches Forschungsinstitut für Segelflug (DFS)	German Research Institute for Sailplanes
Deutsche Versuchsanstalt für Luftfahrt (DVL)	German Test Establishment for Aviation
Dipl.-Ing. (*Diplom Ingenieur*)	Diploma Engineer – academic degree in engineering
Drehlafette	Rotating gun mount
Einheitstriebwerk	Completely Formed Engine Unit
Entwicklungsabteilung	Development Department
Entwurfsbüro	Development Office
Erprobungsstelle(n)	Test Centre(s)
Fernaufklärer	Long-range reconnaissance aircraft
Fernbedienbare Hecklafette	Remotely-controlled rear gun mount
Flugbaumeister	Construction Chief
Flugkapitän	Flight Captain
Flugzeugtypenblatt	Aircraft Type or Designation Sheet (RLM)
Forschungsinstitut für Kraftfahrwesen und Fahrzeugmotoren	Research Institute for Motor Vehicle Technology and Vehicle Engines
Führungsstab	Command Staff (e.g. for *Luftwaffe*)
General der Aufklärungsflieger	General of Reconnaissance Aviation
General der Fliegerausbildung	General of Flying Training
General der Jagdflieger	General of Fighter Aviation
Generalluftzeugmeister	Chief of Procurement
Generalquartiermeister	General Quartermaster
Generalstabs-Ingenieur	Engineer on General Staff
Jagdwaffe	Fighter Arm
Jägerstab	Fighter Staff (emergency committee)
Kriegsmarine	Navy
Lufttransportführer	Air Transport Commander
Luftwaffenführungsstab	Luftwaffe Command Staff
Oberste Heeresleitung	Army High Command
Planungsamt	Planning Office (RLM)
Projektbüro	Project Office
RATO	Rocket Assistance Take-Off (Units)
Reichsverteidigung	Air defence of the Reich
Regia Aeronautica	The Royal Italian Air Force
Ritterkreuz	Knights Cross
Reichsluftministerium (RLM)	German Air Ministry
Seekriegsleitung (SKL)	Naval War Staff
Sonderfernaufklärer	Special long-range reconnaissance aircraft
Sonderkommando	Special command or detachment
Sturmgruppe	lit. 'Assault Group' – heavily armed and armoured fighter Gruppe
Technisches Amt	Technical Office (RLM)
Unternehmen	Operation (as in military operation)
Ural Bomber	Long-range bomber able to strike beyond the Ural Mountains
Verbandsführerschule für Kampfflieger	Unit leader's school for bomber pilots
Versuchsstelle für Höhenflüge (VfH)	Experimental Testing Station for High-Altitude Flight
Versuchsverband Ob.d.L	Experimental unit of the High Command of the Luftwaffe
Vorprojektbüro	Preliminary project department
Wolfsschanze	lit. 'Wolf's Lair' – Hitler's East Prussian field headquarters
Zwilling	'Twin' as in coupled or two-gun configuration

'The United States cannot be invaded. The German Air Force has no planes of sufficient range for transatlantic operations. Even if you don't like us, give us some credit for common-sense and reason.'

Hermann Göring to American journalists, Berlin, 20 July 1940

'I well remember that at Augsburg – it was exactly a year ago – I was shown an aircraft that really called for nothing more than to be put into mass production. It was to fly to the east coast of America and back, from the Azores to the American west coast and also to carry a lot of bombs. I was told so in all seriousness. But in those days I was still so trusting, I half believed it.'

Hermann Göring, March 1943

Megatheria

'...they smashed up the city as a child will shatter its cities of brick and card. Below, they left ruins and blazing conflagrations and heaped and scattered dead... Lower New York was soon a furnace of crimson flames, from which there was no escape...'

H.G.Wells, *The War in the Air,* 1908

In January 1908, eight years before the outbreak of the First World War, the prolific British essayist and best-selling novelist, H.G.Wells, published the first in a series of twelve segments of a new adventure tale in the London newspaper, the *Pall Mall Gazette*. The story was entitled *The War in the Air*. Wells believed that his stories went much further than providing just another 'ripping yarn', and that they served to warn the public about the dangers that he predicted would arise from technological progress and invention – inventions which, in the 'wrong' hands – could threaten the very safety of the British Empire.

The inspiration for *The War in the Air* lay in the development and highly-publicized flights of Ferdinand *Graf* von Zeppelin's giant airships in Germany. In 1907, von Zeppelin's airship, LZ3, had demonstrated its ability to stay aloft for nearly eight hours and traverse a distance of nearly 320 km while carrying a crew and a substantial load of water ballast. The massive craft emerged again from its floating hangar near Friedrichshafen to fly on numerous occasions in 1908 and 1909 when Zeppelin was able to treat an impressed *Kaiser* Wilhelm and Prince Heinrich von Preussen and other influential individuals with flights over Lake Constance.

In England however, the public was becoming increasingly anxious about Germany's ambitions and Wilhelm's plans to build a navy equal to Britain's. Rumours abounded that the Luftschiffbau Zeppelin was planning to construct an even larger airship that would be capable of remaining in the air for 24 hours.

From the furiously busy pen and imaginative mind of H.G.Wells, flowed a story in which the Germans, intent on attacking the United States, despatch an 'air fleet' to operate over the Atlantic. There, the air fleet would destroy the American naval fleet as it fought a hopeless battle against German dreadnoughts, following which the great airships '... of gas and basket-work ... the weirdest, most destructive and wasteful megatheria in the whole history of mechanical invention' would fly at 90 mph to New York before news of the American naval defeat could reach the city. In Wells' story, the local government is panicked by the appearance of such awesome aircraft and surrenders but, much to the

Far left and left In 1908, the prolific British science fiction writer, H.G.Wells, wrote a work in which a fleet of giant German airships – or airborne 'dreadnoughts' – flew across the Atlantic to bomb New York, where they inflicted terrible damage and wreaked fear amongst the populace. To accompany the book, an artist was commissioned to depict the German machines as they attacked the American Navy (**Left**) and left the city in flames (**Far left**). Wells sought inspiration for his story from the formidable and impressive airship designs of Ferdinand Graf von Zeppelin.

infuriation of the German air commander, the city's angry and stubborn residents determine to resist. Faced with no alternative the Germans decide to bomb New York into submission. From aboard the airship *Vaterland*, one of Wells' characters in *The War in the Air* watches the horror below:

> *'As the airships sailed along they smashed up the city as a child will shatter its cities of brick and card. Below, they left ruins and blazing conflagrations and heaped and scattered dead: men, women, and children mixed together as though they had been no more than Moors or Zulus, or Chinese. Lower New York was soon a furnace of crimson flames, from which there was no escape. Cars, railways, ferries, all had ceased, and never a light led the way of the distracted fugitives in that dusky confusion but the light of burning.'*[1]

Thankfully for New York, the fiction of H.G.Wells never became fact in the devastating world war which was to follow in 1914 and yet for the people of London, Wells' writing would prove grimly prophetic. The German *Oberste Heeresleitung* recognised the significant offensive contribution which the airship could make to the conduct of the war and from the first year of conflict, sent its army and naval airships to bomb London and other targets across England. The appearance of these monstrous machines in the skies over England and the increasing levels of loss of life and destruction caused by their bombs generated panic and even riots amongst a civilian population which had always assumed that wars were fought on foreign soil. The German Army had undertaken nearly 200 airship missions over England by 1918 and had dropped some 36,000 kg of ordnance on British targets, which was supplemented by 307,000 kg of ordnance dropped by Navy airships on enemy vessels in the North Sea and on harbours and towns from Cornwall to Scotland. In January 1919, *The Times* newspaper reported that 498 civilians and 58 military personnel had been killed, with another 1,913 injured as a result of the 'Zeppelin' raids.[2]

However, during the First World War and in the immediate post-war years, the United States felt relatively secure and distant from the conflict in Europe. Many Americans believed that geographic isolation offered protection. But there were those who voiced a contrary view. One such voice belonged to Brigadier General William M. 'Billy' Mitchell. Mitchell had served on the US Army General Staff since 1912 and between 1916-1917 had paid for his own flight training, following which he was sent to Europe to observe French and British military air operations. He was subsequently appointed head of the US Army Expeditionary Force's Aviation Service and after the War rose to become Assistant Chief of the new Air

Left The German airship, the L 13, was the most successful Navy airship to operate over Western Europe and the British Isles, recording 159 flights for a total of 69,100 km. Completed at Friedrichshafen in July 1915 it was commanded by Kapitänleutnant Mathy, Kapitänleutnant Eichler, Kapitänleutnant Schwonder and Oberleutnant zur See Fleming. Heinrich Mathy flew 14 airship bombing raids over England, including central London, more than any other German naval airman. One raid on London in L 13 caused £530,000 of damage.

Above Brigadier General William M. 'Billy' Mitchell warned that the development of the military aircraft would change the United States' inherent sense of defence by removing its geographic isolation.

Above The Italian air power theorist, Marshal Guilio Douhet, established an air doctrine which supposed the invincibility of mass formations of heavily-armed heavy bombers deployed against centres of civilian population and industry.

Service. Mitchell believed that the development of the military aircraft fundamentally changed the defensive position of the United States. The aircraft's ability to travel much farther and faster than previous means of transportation removed the isolation the USA had previously counted on as part of its security. Furthermore, Mitchell believed in the creation of an independent organisation responsible for air power and that once an independent air force had been established, its primary mission was to attack and destroy the enemy's air forces.

Later, during the Depression years of the early 1930s, Mitchell's disciples at the Air Corps Tactical School (ACTS) were greatly influenced by the Italian air power theorist, Marshal Guilio Douhet, who prophesied the brilliant and conquering future of the aerial bomber, derived initially from his personal war experiences gained in combat during Italy's war against the Turks in 1911. Translated into English by 1921, Douhet's theories were to transform military aviation doctrines around the globe and many came to believe with near-mysticism in the coming age of air power, envisaging fleets of high-flying bombers so well-armed that they would fend off – and even destroy – an enemy's disorganised and outnumbered fighter forces. In much the same way as H.G.Wells had written in 1908, American writers opined that bombers would be used in mass over urban centres, their vast numbers darkening the sky, destroying factories and troop centres with pinpoint accuracy and that terrorised citizens of major cities would be all too ready to surrender under a rain of falling bombs. Artists were hired to paint scenes of an enemy's open sky dotted with Air Corps bombers raining endless sticks of bombs upon defenceless factories. Enemy fighters were depicted firing harmlessly from out of range of the bombers, or spinning down in flames as victims of their 'fortress-like' armament. It was a fantasy, but not without its advocates.[3] Certainly, the advocates at the ACTS espoused the virtues of bomber technology over the further development of 'pursuit' or fighter interceptor aviation. In 1934, for example, Captain Harold L. George and Captain Robert M. Webster of the ACTS, carried out an in-depth analysis of the vulnerability of New York City to daylight precision bombing. They came to the conclusion that if bombs could be used to strike essential services – i.e. water, electricity and transportation – the effect would be to make the city 'unliveable'.[4]

Meanwhile in Europe, following the end of the French occupation of the Ruhr, Germany was experiencing the beginnings of an economic recovery and internal stability for the first time in over a decade. Despite soaring inflation and widescale unemployment affecting much of the population, her currency had been restored and foreign investors were starting to return. Furthermore, the German industrial 'machine' was producing more coal and steel than it had done before 1914. By 1930, despite her industry actually operating at between only 50 and 80 per cent of its capacity, Germany had become world-leader in the export of finished products and ranked second after the United States among the world's exporting countries.[5]

It was therefore with some confidence that the Germans launched the 1932 *Deutsche Luftsport-Ausstellung* – the German Aero-Sport Exhibition – in Berlin. It was here that the Augsburg-based aircraft designer, Willy Messerschmitt, presented the international aeronautical fraternity with a model of a radical new design for a long-range aircraft. Known as the M 34, Messerschmitt had worked on the project with his brother-in-law, *Professor* Georg Madelung, who had designed a world record-holding sailplane in the early 1920s. Messerschmitt's ambition was to develop an aircraft which incorporated a number of key aerodynamic refinements that had appeared in his earlier designs and which would be powered by the technology of newly available diesel engines. The M 34 was to be made up of a slim fuselage, a wing similar

Below Willy Messerschmitt (centre) in discussion with Otto Strohmeyer (left), the Chairman of the Board of the Bayerische Flugzeugwerke AG from 1931-32, and Theo Croneiss, a supporter of Messerschmitt in his early years of business. In the background is a BFW U 12 Flamingo.

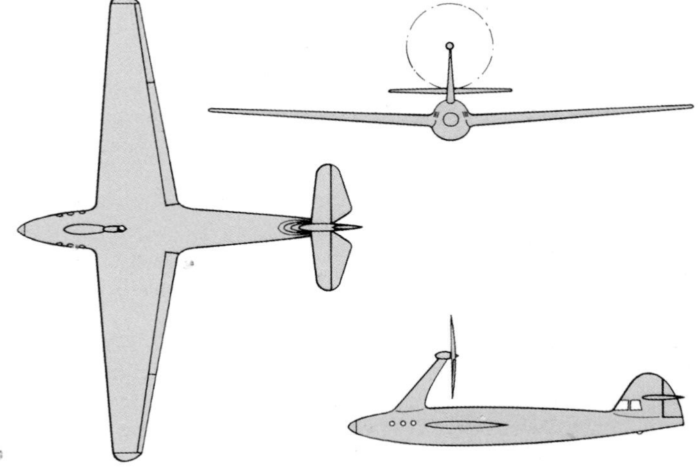

Below right Willy Messerschmitt's M 34 project of 1932, in which he endeavoured to design an aircraft which would possess a range of 20,000 km and be able to reach the Antipodes from Germany or to fly from any point of the globe to another without refuelling. Though the project did not progress beyond the drawing board, it was nevertheless one of Messerschmitt's earliest attempts at tackling the challenge of range.

to that of a sailplane with a 34 metre span, a cockpit built into the vertical stabilizer at the aft of the fuselage and retractable landing gear. With a single 800 hp engine, or two rated at 400 hp each, Messerschmitt projected that such an aircraft would possess a range of 20,000 km and would thus be able to reach the Antipodes from Germany or to fly from any point of the globe to another without refuelling.[6]

It was a dream which the aircraft designer cherished and would hold on to for years to come. However, the reality was that Messerschmitt lacked sufficient development funds to proceed, ensuring that nothing further came of the design and while Germany enjoyed its economic rejuvenation, Messerschmitt was struggling with financial difficulties.

During 1929, one of Messerschmitt's companies, the *Bayerische Flugzeugwerke AG* (BFW or Bavarian Aircraft Factory), had drawn up license agreements with the Baltic states of Lithuania, Latvia and Estonia and the Eastern Aircraft Corporation of Pawtucket, Rhode Island. The American company was to build the M 18 and M 26 airliners, but the Wall Street crash of 1929 meant that the agreement had to be cancelled after Messerschmitt had received a single payment of US $189,000. The US stock market crash created severe economic problems which rapidly reverberated around the globe. Not only did the deteriorating international economic situation affect German aircraft manufacturers such as Arado, BFW, Dornier, Heinkel and Junkers, but it also aided the growth of National Socialism (which until then had only manifested itself as a minor political party) and the rise of Hitler, who promised to redress the nation's economic difficulties.

Above Erhard Milch, the Managing Director of Deutsche Luft Hansa, first clashed with Messerschmitt in 1931 when two of the aircraft designer's M 20 airliners crashed killing eight officers. Milch subsequently set about discrediting Messerschmitt in any way he could, and an enmity was to grow between the two men that lasted for the rest of their lives and would prove detrimental to the efficient functioning of the German aviation industry during the Second World War.

By the end of 1930, BFW had losses in the region of 600,000 RM, and even an injection of 250,000 RM by the Stromeyer-Raulino family, which on 1 July 1928 had bought out the shares held in the company by the *Reichsverkehrsministerium*, failed to improve matters. Then, during the spring of 1931, two of *Deutsche Luft Hansa's* BFW-built M 20 airliners crashed in unexplained circumstances. In one of the crashes, the eight passengers, all *Reichswehr* officers, were killed and the accident received widespread adverse publicity in the German national press and was even discussed in the *Reichstag*. At that time, the managing director of *Deutsche Luft Hansa* was Erhard Milch, a 39 year-old former First World War artillery officer turned fighter pilot. One of the officers killed in the M 20 crashes was a long-time personal friend and assistant to Milch.[7] Milch alleged that Willy Messerschmitt had not designed the M 20 with the necessary safety measures as a result of his tendency to save weight at all costs. Although Milch's allegations remained unproven, the latter promptly cancelled *Deutsche Luft Hansa's* outstanding orders for the M 20 and demanded the repayment of all deposits the airline had paid to BFW. These demands could not be met by BFW, for manufacture of the remaining M 20s had advanced too far to accept cancellation. This was to prove the final blow in BFW's financial crisis, and it was forced to register itself bankrupt on 1 June 1931. Messerschmitt's other company, the *Messerschmitt Flugzeugbau GmbH*, which still retained a large degree of autonomy, was able to carry on, and eventually *Deutsche Luft Hansa* was persuaded to honour its agreement to buy the M 20 and M 28 aircraft. Understandably, the dispute between Messerschmitt and Milch was not helped by the fact that Messerschmitt happened to be a personal friend of Theo Croneiss, owner of *Nordbayerische Verkehrsflug*, which was *Deutsche Luft Hansa's* only serious competitor. The enmity between Messerschmitt and Milch continued to the end of the Second World War and overshadowed many Messerschmitt projects including the Me 264.

An agreement was made with BFW's creditors in December 1932, allowing BFW and its workforce of 82 personnel to recommence trading on 1 May 1933. The following month, Milch told Friedrich Seiler, a banker who had joined the board of BFW and who had offered to underwrite a part of its financing: 'You are bailing out a would-be industrialist who will never make the grade. And the moment you are down, Messerschmitt won't help you up. He'll kick you in the teeth!'[8] The company also had to fight against adverse publicity, such as when the British aviation yearbook, *Jane's All the World's Aircraft* reported: 'The BFW company has found itself in difficulties. For the time being therefore, its affairs are being handled by the *Messerschmitt Flugzeugbau GmbH* of the same address.'[9]

Above General Hermann Göring, the head of the new Reichsluftfahrtministerium. There was a mutual admiration between Göring and Milch, and though he at first declined Göring's offer to become State Secretary for Aviation, Milch eventually accepted when Göring used Hitler to persuade him.

Shortly after Adolf Hitler came to power in January 1933, a German Air Ministry (*Reichsluftfahrtministerium* or RLM) was established on 27 April. The new ministry was placed under the leadership of *General* Hermann Göring with Erhard Milch, in the post of State Secretary, responsible for procurement. Almost immediately most German aircraft companies were awarded huge contracts and financial help for their expansion, but not BFW. Its sole order was a contract for the license-building of 12 He 45 C reconnaissance aircraft. When Seiler and Croneiss visited Milch privately to offer proposals to save the business, Milch simply brushed these aside. The newly-appointed State Secretary told his visitors that, in his opinion, although Willy Messerschmitt was an intelligent and capable designer, he was also egotistic, ungrateful and untrustworthy. Furthermore he had shown remarkable insensitivity in not paying his respects to the victims of his own aircraft's crash.[10]

The lack of orders led the company's Greek-Cypriot commercial director, Rakan Kokothaki – who in 1931 had gone so far as to sell his car to keep Messerschmitt's beleaguered business afloat – to travel to Rumania in June 1933, where he gained an order for the design of the M 36, a small airliner capable of carrying six passengers.[11] The fact that BFW had sought a foreign development contract incurred the wrath of *Oberstleutnant* Wilhelm Wimmer, then head of the RLM's *Technisches Amt* (Technical Office), who was considered to have '... the best technical mind in the *Luftwaffe*.'[12] Wimmer argued that German aircraft manufacturers should not be working for foreign countries at a time of such rapid expansion and re-armament. Messerschmitt countered by saying that since no German-originated contracts had been forthcoming, he was forced to seek work abroad. As a direct result of this, the RLM gave BFW an order to design an aircraft to take part in the 4th *Challenge de Tourisme Internationale* of 1934, and also allowed it to

Above Willy Messerschmitt (right) studies plans with two of his most loyal executives. During the early 1930s, Rakan Kokothaki (left) went to great efforts to keep BFW in business. From 1942, he would act as Operations Manager for Messerschmitt A.G., taking over the role from Theo Croneiss. Fritz Hentzen (centre) joined BFW from Fokker and oversaw the production of all Messerschmitt's series aircraft, including the Me 264.

enter the competition for the design of a new fighter aircraft. Despite being permitted to enter the fighter competition, BFW was considered to be a rank outsider by the Technical Office. It was felt that Messerschmitt possessed no experience in the design of high-speed combat aircraft, and thus had little chance of winning against the more experienced Arado and Heinkel companies.

At some point in early 1934, an event occurred that almost changed the entire course of Willy Messerschmitt's career and the future of BFW. The University of Danzig offered him the post of Professor of Aeronautics, acceptance of which would have released Messerschmitt from the many difficulties he had experienced in trying to establish himself as a leading member of the German aviation industry. Yet despite his undeniable successes in the field of transport and sporting aircraft design, Willy Messerschmitt felt he remained out of favour with the RLM who viewed his abilities with some suspicion. His initial impulse was to resign his position with BFW and accept the offer of the professorship which he found personally both attractive and flattering. In order to determine the extent of any prejudice the RLM's Technical Office might hold against him, Messerschmitt advised Wimmer of the offer and inquired if any importance was attached to his work in the future. Messerschmitt was promptly notified that no importance was attached to his person for the subsequent development of German aviation and that he would do well to accept the professorship at Danzig! Fortunately however, Messerschmitt's friends (who included Rudolf Hess, the *Führer's* deputy) and supporters convinced him that he did have a future as an aircraft designer in Germany and, much to the annoyance of State Secretary Milch, he declined the professorship. Although, he was unaware of it at the time, Willy Messerschmitt's fortunes were about to improve for the better as work commenced on the fighter competition contract, which was eventually to result in the most famous and successful aircraft that ever bore his name, the Bf (later Me) 109.[13]

The same year that Willy Messerschmitt began to work on the design of the Bf 109, *Oberst* Walther Wever, the fledgling *Luftwaffe's* first Chief of Staff, supported by a number of his key officers such as Wimmer and Wolfram *Freiherr* von Richthofen, Wimmer's assistant, decided that the *Luftwaffe* would need a heavy, multi-engined strategic bomber.[14]

Despite a long-held popular view to the contrary, Germany remained committed to the concept of building a strategic bomber fleet from 1933 through to the mid-war years, and Wever played a significant role in fostering this ambition. The air power historian, James S. Corum, has described Wever as playing '... an important, even a central part in the development of air power thought in Germany, and his work would significantly impact the development of the *Luftwaffe*... [he] came to be regarded as the dominant personality of that service.'[15]

Above Oberst Walter Wever, the first Chief of Staff of the Luftwaffe. A hard-working, and popular leader, he advocated the development of a so-called 'Ural Bomber' which would be able to reach targets deep in eastern Russia, the country which he believed would become Germany's primary enemy in any future armed conflict. His policies were abandoned upon his premature death in a flying accident in June 1936.

Wever, a man of charm and considerable, far-sighted intellect but with a reputation as a hard-working and demanding commander, did not believe that bombing alone could win a war, nor did he support many of Douhet's theories. However he did believe that Germany's main enemy in the future would be Russia and not France. Therefore, any future war would involve industrial targets deep within Russia, perhaps even east of the Ural mountains, hence the concept of an ultra-long-range bomber – or '*Ural Bomber*' – was born. Encouraged by Wimmer, Wever ordered a specification to be issued by the RLM *Technisches Amt* to Junkers and Dornier for an aircraft of this type. It was a specification that was well in advance of the plans of most of the world's air forces and it would stretch the resources of the still very young post-war German aircraft industry to its limits – demanding the carrying of a heavy bomb load over a great distance with the benefit of speed and good defensive armament. By May 1934, the development of such a heavy bomber became the *Luftwaffe's* top priority.

Junkers subsequently produced one prototype of the big, all-metal, four-engined Ju 89 by 1936 and from Dornier emerged three Do 19 prototypes. The Dornier was a large four-engined machine with a thick, low mid-wing and braced vertical stabilizers mounted on top of the tailplane. It was designed originally to carry no fewer than six gunners. Both designs were underpowered for their size and weight, particularly the Do 19 which was to be fitted initially with nine-cylinder, 715 hp air-cooled Bramo engines, generating a maximum speed of only 370 kph over a maximum range of 1,995 km with a bomb load of 1,600 kg. The first prototype, the Do 19 V 1, first flew on 28 October 1936 with the Bramo powerplants, while the V 2 was enhanced by the installation of more powerful BMW 132F nine-cylinder radials. The V 3 was to have carried the planned armament of two 20 mm MG FF cannon housed in two two-man hydraulically-powered turrets and two MG 15 machine guns in nose and tail positions. However, the V 2 and V 3 were scrapped prior to their completion.

The unarmed Ju 89 V 1, powered by four Jumo 211A 12-cylinder liquid-cooled engines, made its first flight two months after the Do 19, and was followed in early 1937 by the V 2, fitted with 960 hp Daimler-Benz DB-600A liquid-cooled engines. The aircraft could reach 390 km/h at maximum speed and could carry a slightly heavier bomb load than the Dornier over a marginally greater range, together with a crew of nine, including five gunners operating a similar armament to that carried by the Do 19. But, comparatively, neither aircraft could match the performance of the new American B-17 Flying Fortress.

Frustrated and disappointed by the design of the new bombers even before they first flew, on 17 April 1936, with Wever's approval, Wimmer issued revised specifications for a heavy bomber and demanded totally new designs. Then disaster struck when on 3 June 1936, Wever was killed flying a Heinkel He 70 as he left Dresden airfield en route for Berlin. His loss was not easy to replace.

His eventual successor was the extremely capable Albert Kesselring, who was pulled from running the RLM's Administrative Office where he had impressed Göring with his accomplishments during the initial phase of establishing the new *Luftwaffe*. Popular with those who worked for him, he brought energy and considerable organisational insight into his new position.

Above Oberst Ernst Udet was appointed Chief of the Technisches Amt (Technical Office) of the RLM on 9 June 1936. A charming and extrovert 'pilot's pilot' and holder of the Pour le Mérite, he was an advocate of small, fast aircraft and had little understanding of Wever's earlier plans for a long-range, heavy bomber.

But with *Generaloberst* Göring consolidating his power-hold over the senior command of the *Luftwaffe*, so he began to favour those whom he considered pliable and as a result the competent but steadfast Wimmer was sidelined in favour of *Oberst* Ernst Udet. Udet was an old comrade of Göring's, a flamboyant and famous fighter pilot who had been awarded the *Pour le Mérite* during the First World War and who had ended that conflict with 62 aerial victories. He was also an advocate and enthusiast of small, fast aircraft and in 1934 had visited the United States where he had been favourably impressed by the Curtiss Hawk dive-bomber then in service with the US Navy. As a pilot first and foremost, rather than an air power strategist, he had little interest for or appetite in the prospect of large, long-range heavy bombers, nor did he understand Wever's reasoning or his commitment to such a type of aircraft. Rather, Udet supported the notion of tactical precision bombing of the kind the new generation of dive-bombers could undertake. In this, fundamentally, Kesselring shared his view and further considered that a two-engined aircraft would suffice for any war envisaged in western Europe.[16]

Simultaneously, Milch was convinced that production of a long-range, four-engined bomber would prove too costly in metal and other vital raw materials. The fate of the Junkers and Dornier bombers was thus sealed when on 29 April 1937, despite protests from the *Technisches Amt*, Göring gave the official order for all further development work on the '*Ural Bomber*' to cease, and the project was shelved. 'The *Führer* will never ask me how big our bombers are,' Göring commented, 'but how many we have.'

Yet that same year, at a meeting with Göring and Milch, the Chief of Branch 1 of the *Luftwaffe* Operations Staff, *Major* Paul Deichmann, pressed Göring for authorisation to continue work on a four-engined bomber '... at all costs.' As Deichmann recalled in a post-war paper: 'I emphasised that the four-engined bomber was capable of a far greater flight range than could ever be developed for its twin-engined counterpart. I continued explaining that a four-engined bomber was capable of attaining a sufficiently high service altitude to keep it safely out of the range of anti-aircraft artillery fire; its considerably greater carrying capacity would permit it not only to carry a greater number of bombs, but also heavier armour plating and more and better airborne armament. Its higher speed would help reduce its vulnerability to attack by enemy fighter aircraft.

'General Milch interrupted, demanding to know from where I had obtained all this information on the "fantastic" performance of the four-engined bomber. He told me that his own aeronautical engineers had come up with far less favourable prognoses. I replied that I was aware of the views held by this particular group of engineers within the RLM, but that a considerable number of the engineers connected with the *Technisches Amt* were of a quite different opinion. The only way to determine which group was right was to let the developmental work continue. At this Milch declared that all available industrial capacity was needed for the production of Ju 88s...

'In summary, General Milch pointed out the following facts: 1) the much vaunted advantages of the four-engine bomber were far overrated, both in Germany and abroad; 2) what would be the point of its being able to fly at 52,800 metres? 3) our industrial capacity would permit a fleet of only 1,000 heavy bombers, whereas several times that many twin-engined bombers could be produced; 4) the development of a four-engined bomber, even for limited production as test models, would endanger the Ju 88 programme...

'I begged Göring not to decide against the four-engined bomber without further evidence, but to let the developmental work on it continue. Despite my pleas, Göring determined that work on the four-engined bomber should be dropped in as much as it might interfere with successful accomplishment of the Ju 88 programme.'[17]

It seems however that Deichmann's pleas did have some effect. In a somewhat surprising 'U-turn' later in 1937, Milch quietly authorised specifications to be drawn up for a 'multi-functional aircraft' which could deliver a 5-ton bomb load to New York and a lesser tonnage to targets in the central United States, whilst also being capable of conducting reconnaissance missions as far as the American West Coast.

Above Major Paul Deichmann pressed Göring for authorisation to continue work on a four-engined bomber '... at all costs', but he was ignored

This page During 1934 the Luftwaffe Chief of Staff, General Walter Wever, proposed the construction of a long-range strategic bomber able to attack targets in the north of Scotland or in the Urals from German bases. The specification for such an aircraft was placed with the Dornier and Junkers companies, both of whom had experience with long-range multi-engined aircraft. The Dornier project, the Do 19, was a somewhat ugly mid-wing monoplane with slab-sided fuselage and twin fins and rudders. It was powered by four 715 hp Bramo 322 H-2 radials and had a retractable undercarriage. A crew of nine was to be carried, comprising pilot, co-pilot/navigator, bomb-aimer, radio operator and five gunners. Armament was to comprise a nose turret housing a 7.9 mm MG 15 machine gun, a similar gun in the tail and two large two-man turrets in both the dorsal and ventral positions each housing a single 20 mm cannon. The prototype Do 19 V1 first flew on 28 October 1936 without armament. It was initially unmarked, but the civil registration D-AGAI was later applied. Tests showed that the aircraft was already underpowered, managing only to attain a speed of 315 km/h. It was quickly realised that the heavy and cumbersome two-man turrets would further reduce the aircraft's effectiveness and these were abandoned. Nevertheless, work went ahead on the 810 hp BMW 132 F-powered Do 19 V 2 and the V 3 the second of which was to be the first with armament. However, the death of Wever in an air crash in June 1936 had forced a rethink of the original 'Ural Bomber' proposal and the type was abandoned and all existing prototypes were scrapped.

Below The Do 19 V 2 was virtually complete when, like the half-finished V 3, it was scrapped. The V 1 prototype was experimentally fitted with a mock-up glazed gondola fairing beneath the nose which was to house the bomb-aimer.

Left A busy scene at the Junkers airfield at Dessau as the Ju 89 V 1 is prepared for a test. These tests were to show that the aircraft had a maximum speed of 390 km/h, but installation of armament would have reduced this performance considerably. It was similar to the Do 19 with a 7.9 mm machine gun in the nose and tail and a 20 mm cannon in a dorsal and ventral turret. The aircraft in the background of this photograph is the Ju 88 V 1 which was also undergoing tests at this time.

Right The Ju 89 was designed to the same 'Ural Bomber' specification as the Do 19 but proved to have a much better performance than its rival. Basically an enlarged Ju 86, the prototype Ju 89 V 1 was powered by four 1,075 hp Jumo 211 engines driving three-bladed Hamilton propellers. It made its first flight in December 1936, later having the registration D-AFIT applied.

Below The Ju 89 V 1 was joined by a second prototype early in 1937. This was generally similar to the V 1, but was powered by four Daimler-Benz DB 600 engines and was later registered D-ALAT. Construction of a third prototype, which was to carry mock-ups of the two fuselage turrets, began but following the cancellation of the original 'Ural Bomber' programme on 29 April 1937, the airframe was converted to serve as the forerunner of the Ju 90 commercial transport. Both the Ju 89 V 1 and V 2 served briefly with the transport unit KGrzbV 105 during the invasion of Norway in April 1940.

'The Banana Plane'

'I completely lack the bombers capable of round-trip flights to New York with a 5-ton bomb load. I would be extremely happy to possess such a bomber which would at last stuff the mouth of arrogance across the sea.'

***Generalfeldmarschall* Hermann Göring, 8 July 1938**

For Willy Messerschmitt, the year 1937 saw events take a dramatic change for the better. Firstly, his credentials were favourably elevated when he received an honorary Professorship from Munich Technical High School, where he had obtained his engineering degree some fourteen years earlier. On 1 April, he was appointed a member of the newly-formed *Deutsche Akademie der Luftfahrtforschung* (German Academy for Aviation Research). It was a high honour to belong to the Academy, and a clear indication that finally, after his many tribulations, Messerschmitt had been accepted by the aviation establishment in Germany.

Then, between 23 July and 1 August, six of his Bf 109 prototypes were included in the German team which participated in the 4th International Flying Meeting held at Zurich-Dübendorf in Switzerland. This was the first time that the aircraft had been shown to the public, apart from a brief appearance at the 1936 Olympic Games held in Berlin. The Bf 109 proved to be an outstanding success and won four first prizes: the climb and dive competition, the speed event around a 50 km closed circuit, the *Alpenflug* and the team *Alpenflug*. The competing foreign teams were completely outclassed by the Bf 109's outstanding qualities, and its impact came as something of an eye-opener to the British and French Air Forces. Considerable expansion also took place within the company during 1937. New factory hangars were built and others enlarged and a new purpose-built administration and design building was completed. Within a year, the Regensburg factory had been completed, and work was under way on the Bf 108 and the new twin-engined Bf 110 fighter.

On 11 November BFW's international prestige was increased still further, when the Bf 109 V 13, powered by a specially-boosted Daimler-Benz 601 engine, achieved an average speed of 610.95 km/h over a 3 km course, and thus established a new World Speed Record for Landplanes – the first time this much coveted record had been won by Germany.

This event was cause enough to bring none other than Adolf Hitler on an official visit to the Messerschmitt factory at Augsburg eleven days later. Accompanying the *Führer* were Milch and Udet who headed a delegation of officers and engineers from the RLM *Technisches Amt*. As Hitler toured the works,

Messerschmitt approached him in one of the assembly halls and, signalling for two great hangar doors to be opened, revealed the mock-up of a very large four-engined aircraft. This was probably a strategic bomber project known as the 'Bf 165' which Messerschmitt viewed as a contender to the Do 19 and Ju 89. The Bf 165 was designed with a long, tapering tail boom and considerable defensive armament. In his memoirs, Hitler's *Luftwaffe* adjutant, Nicolaus von Below, records the moment the doors rolled back: '*At this the RLM gentlemen became very uneasy and the horror on Milch's face was evident.*'

Hitler listened carefully as Messerschmitt explained what the mock-up represented – a potential four-engined, long-range bomber, able to fly 6,000 km and intended to carry a one-tonne bomb load at 600 km/h. Milch and the RLM officials quietly and politely scoffed at such a suggestion, but Hitler – who understood relatively little about aircraft – said little other than making the observation that if a fighter could manage 600 km/h, then a bomber would need 650 km/h, meaning that armour and weapons would have to be sacrificed for speed. Messerschmitt responded respectfully that high speed was not possible at that stage because of insufficient power from the engine types presently available.

On 22 November 1937, Adolf Hitler visited the Messerschmitt factory at Augsburg in recognition of the Bf 109 V 13's record-breaking speed performance above the Augsburg-Buchloe railway line eleven days earlier. As Hitler toured the works, Professor Messerschmitt (right) signalled for two hangar doors to be opened and presented the Führer with what was probably the mock-up of the Bf 165, a very large, four-engined, strategic bomber project. This was the last thing Udet – seen next to Messerschmitt – wanted Hitler to see. Hitler, it seems, much to Messerschmitt's frustration and Udet's relief, remained indifferent. To Hitler's right is Fritz Hentzen, Messerschmitt's Production Manager.

Hitler shrugged; it was only correct that the *Luftwaffe* was subject to limits on raw materials. In the *Führer's* view – which was influenced by Göring's desire to offer encouraging production figures based on the Ju 88 – a twin-engined fast bomber was preferable. Hitler moved on and the subject was closed.[1]

Messerschmitt, however, was not about to be deterred by this incident; the previous year his company had allocated 100,000 RM towards the mock-up of the Bf 165 and he still harboured ambitions to design and build a long-range aircraft which would be able to fly distances up to 20,000 km.[2] His short-term aim was to design a high-speed transport for economically hauling high-value perishables from the tropics and hence his project engineers christened his proposal '*Das Bananenflugzeug*' ('The Banana Plane').[3] His commitment to the concept was rewarded at the beginning of 1938, when the RLM issued new specifications to the aircraft manufacturers. These contained guidelines for a future long-range bomber programme. Operational altitude was set at 5,000 m, and range at 6,700 km with a bomb load of one tonne and a crew of four.[4]

In fact Messerschmitt was already working on more than one long-distance design. One such project – known as the P 1062 – was planned initially as a prestigious, record-breaking aircraft which was to have carried the Olympic torch non-stop from Berlin to Tokyo for the Olympic Games of 1940. This held appeal for Hitler and the aircraft subsequently became known unofficially as the '*Adolfine*' and in some quarters the '*Führerflugzeug*'. It was eventually built in prototype form as the Messerschmitt Me 261. The all-metal aircraft was to have been fitted with two 2,700 hp Daimler-Benz DB 606 A/B 24-cylinder engines each of which was formed by coupling two DB 601 12-cylinder, liquid-cooled engines side-by-side and connecting them to a single shaft which drove a four-bladed airscrew. The power offered by this arrangement was needed in order to carry the large volume of fuel contained in the aircraft's wing tanks. A range of 11,000 km was attainable at economic cruising speed. Although payload would be limited, the Me 261 was to have had a crew of five with bunks housed in its narrow fuselage. From a military perspective, the RLM saw the aircraft having value as a courier machine with accommodation for eight passengers, a transport, a bomber and long-range reconnaissance machine fitted with two automatic cameras for missions over the Atlantic. A specification was issued to install bomb-release gear and defensive armament in the second prototype. Such measures proved impossible however, since the inclusion of guns would have reduced fuel allowance to a level where the aircraft held no significant advantage over other types. The third and final prototype, the Me 261 V 3, was completed in 1943 and redesigned to accommodate a crew of seven and was powered by a pair of 2,950 hp 24-cylinder DB 610s, each comprising two coupled DB 605s.

Simultaneous to working on the Me 261, under *Professor* Messerschmitt's personal instruction, his company's *Entwicklungsabteilung* (Development Department) at Augsburg commenced work on another, larger long-range project – the P 1061. This design incorporated four single engines and, in returning to his '*Bananenflugzeug*' project, Messerschmitt stipulated that range should be in the area of 20,000 km.[5]

But despite continued interest in a long-range bomber from the *Technisches Amt*, objections from the high-level medium-bomber advocates within the RLM did not subside. One day in 1938, Udet apparently remarked to Ernst Heinkel in a corner of the aircraft builder's conservatory: 'In future there won't be any more multi-engined bombers unless they can attack as dive-bombers. By its accuracy, a medium-sized, twin-engined machine, which, in a dive, can hit the target with its bomb load of 900 kg, has the same effect as a four-engined giant which carries 2,700-3,600 kg of bombs in horizontal flight and can only drop them inaccurately. We do not want these expensive, heavy machines which eat up more in material than a medium, twin-engined dive-bomber costs. Junkers has completed his first twin-engined *Stuka*, the Ju 88. We can build two or three with the same amount of material that a four-engined machine needs and achieve the same bombing effect. Jeschonnek is absolutely delighted. Furthermore, with the cheap super-*Stukas*, we can build up the numbers the *Führer* wants.'[6]

Göring, Udet and Kesselring based their opposition to the notion of a four-engined bomber on the drain it would cause to resources of raw materials. The decision to go all-out in favour of the Ju 88 was taken not as a result of Hitler's belief in *Blitzkrieg*, but because the leadership mistakenly thought that in the Ju 88 it had the strategic means, whatever the scope of Hitler's ambitions. Therefore, strategic *thinking* had not been lost, but Germany *thought* that it still had a strategic air force, which could – with the Ju 88 – achieve

Right The Me 261 V 1, BJ+CP, made its first flight on 23 December 1940. In this aircraft, Messerschmitt wanted to create a powerful, long-distance, multi-role machine with a range of 11,000 km. Design and fuel accommodation restrictions served to thwart his plans and only three prototypes were ever completed.

Above The Achilles heel of the Me 261 was its undercarriage which suffered from frequent hydraulics problems as well as weak struts and problematic gear doors. Here, mechanics work on an undercarriage leg of the Me 261 V 3, BJ+CR, following a landing at Lechfeld on 16 April 1943. The machine had just flown for almost 10 hours, for 4,500 km over southern Germany and Austria, accomplishing Willy Messerschmitt's goal of long-range, long-endurance flying. The aircraft had been flown by Karl Baur with Gerhard Caroli as flight engineer, the same team later involved in flight-testing the Me 264. Unfortunately, due to a hydraulic failure in the right undercarriage leg, Baur was forced to land the V 3 on its left undercarriage leg while the tanks still had 5,000 litres of the aircraft's 25,000-litre fuel load remaining. The Me 261 leaned to the right, damaging the outer wing section and the propeller blades of the starboard DB 606 engine. Here the aircraft has been propped up on a trestle trolley ahead of repairs.

Hitler's war aims of embarking on short campaigns or major wars against Britain or Russia. As the historian Richard Overy points out: 'It was Hitler, with his tenuous grasp of the problem of air warfare, his high expectation of the Ju 88, fed by Göring's boasting, and his mistaken appreciation of the state of air preparation, who created the war situation for which the *Luftwaffe* was particularly ill-equipped. Germany possessed the largest fleet of bombers in Europe with the most modern equipment and more battle experience than any other air force.'[7]

On 8 July 1938, during a conference of aircraft manufacturers, *Generalfeldmarschall* Göring was questioned as to the prospect of war with the United States – or at least the ability to wage it from the air. He pronounced: 'I completely lack the bombers capable of round-trip flights to New York with a 5-ton bomb load. I would be extremely happy to possess such a bomber which would at last stuff the mouth of arrogance across the sea.'[8]

'Arrogance' may have been a somewhat extreme description of the American position, although officers on the ACTS Staff were so blindly convinced of the superiority and impregnability of the B-17 Flying Fortress four-engined bomber, that this aircraft was seen by them as the ideal contender for the parent Army as its coastal *defence* weapon. If nothing else, a state of complacency continued to exist within the United States throughout most of the 1930s at the expense of air defence, until in September 1938, President Franklin D. Roosevelt authorized an unprecedented increase in aircraft production – to a figure of 10,000 including fighters. This was a timely move, for just one month earlier – and, ironically, a month after Göring's comments to his aircraft manufacturers – an event had taken place which forced many leading politicians and senior military figures in the United States to reassess and abandon their comforting sense of isolation. At 19.59 hrs on 10 August, the four-engined Focke-Wulf Fw 200 S-1 *Condor* airliner, D-ACON, took off from Berlin-Staaken en route for New York. Piloted by the experienced *Flugkapitän* Alfred Henke of *Lufthansa* and his co-pilot, *Hauptmann* Rudolf von Moreau of the *Luftwaffe Technisches Amt*, the aircraft made its course over Hamburg then across the North Sea to Scotland, passing Glasgow and out over the Atlantic at a height of 2000 metres. At 13.40 hrs, the crew sighted the coast of Newfoundland and flew on over St Johns. At 20.55 hrs, the Focke-Wulf landed safely at Floyd Bennett Field in New York having flown non-stop for 24 hours, 56 minutes – a record-breaking flight of 6,370 km. But as the aircraft taxied to a halt outside the US Navy hangar, and excited technicians and reporters crowded around to welcome the jubilant German crew amidst the whir of movie cameras and the pop and flash of camera bulbs, alarm bells began to ring within the highest echelons of the US Army Air Corps. Certainly, General H.H. Arnold, the recently-appointed Chief of the Air Corps, recognised that the *Condor* could be easily converted into a long-range bomber.

Shortly after landing Henke and von Moreau made a transatlantic telephone call to a jubilant Ernst Udet at the RLM in Berlin. Temporarily it seems, as German aviation basked in record-breaking glory, Udet had forgotten his aversion to four-engined, long-range aircraft.[9]

The following year in Germany, as ominous clouds of political discord were gathering over Europe, the RLM asked Junkers, Focke-Wulf and Messerschmitt for proposals for a four-engined bomber capable of attacking the United States from airfields inside the Reich – a so-called '*Amerika Bomber*'. Despite the fact that the RLM had decided that Messerschmitt should design and build only fighters, the firm submitted its new Me 264 – a proposal based on the P 1061. By this time, *Professor* Messerschmitt's influence was such that he was not only authorised to proceed with the design, but was given an immediate order for three such machines.[10] The '*Bananenflugzeug*' was to carry bombs.

Left The Focke-Wulf Fw 200 S-1 Condor, D-ACON, banks low over Floyd Bennett Field, New York, having flown non-stop for nearly 25 hours from Berlin on 11 August 1938. It was a propaganda coup for the Nazis and the resurgent German aviation industry.

Below Followed by the lens of a movie camera, Fw 200 S-1, D-ACON, taxies to a stop outside the hangar at Floyd Bennett Field, New York. Behind the euphoria of the long-distance flight, there were those at the highest level of the US military who were concerned at what the Condor had accomplished and the possibility that it could be converted into a bomber.

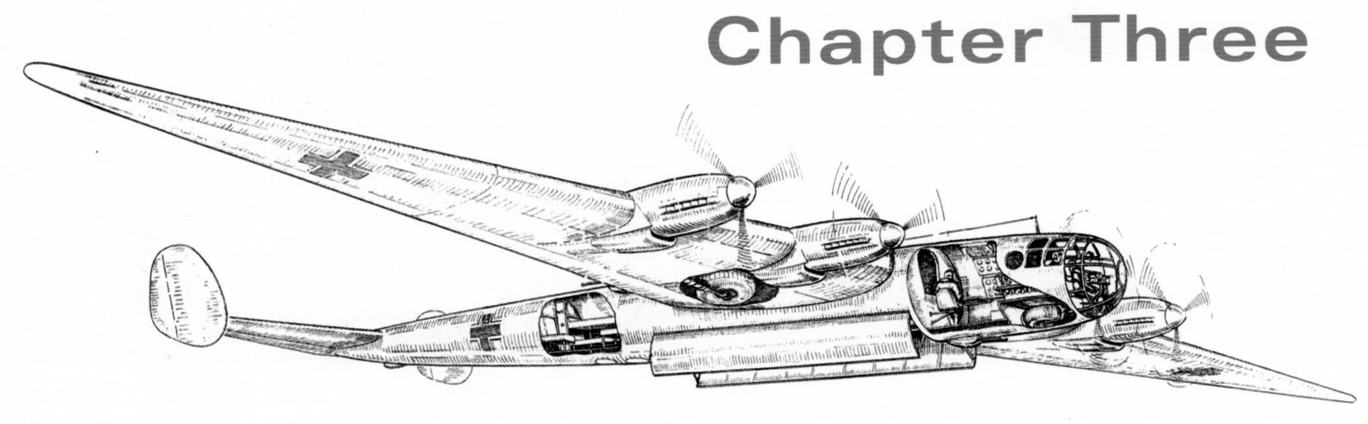

Chapter Three

'Amerika Bomber'

'The task of undertaking nuisance raids against the American East Coast demands so much from the design of the aircraft that it appears impossible at the moment to find a solution without an intermediate development stage. Therefore, the Me 264 represents a vital intermediary stage in development...'

***Generalmajor* Carl-August *Freiherr* von Gablenz, April 1942**

On 18 June 1940, following his swift *Blitzkrieg* conquests in Western Europe and Scandinavia, Adolf Hitler met with his ally, Benito Mussolini, in Munich. The *Führer* was in a bullish mood: the British Expeditionary Force had been driven off the beaches of Dunkirk and his victorious troops had taken Paris only days before. To Mussolini's surprise however, Hitler spoke of his desire to leave the British Empire intact in order to maintain 'world equilibrium' and yet he boasted to the *Duce* that by the end of 1941, Germany would have a fleet of new Messerschmitt heavy bombers ready to attack the USA.

Two days later, on 20 June, while on his way to attend the armistice negotiations with the French at Compiègne, the German leader met with *Admiral* Erich Raeder, the Commander-in-Chief of the *Kriegsmarine*. It seems that Hitler and Raeder discussed the possibilities and practicalities of long-range warfare – or at least the benefits of seizing territory on the very geographic limits of Germany's military reach. Hitler perceived one benefit to this, for he explained to Raeder that having all but taken France, its Jewish population, together with the Jews of Poland, could be transported to the French island of Madagascar in the Indian Ocean.[1]

Supported by Göring, Raeder politely endeavoured to steer the conversation along more practical lines; many senior German commanders understood that there was a distinct possibility of the United States entering the war and Raeder urged the *Führer* to occupy French West Africa, followed quickly by the Atlantic islands of Iceland, the Azores and the Canaries before the Americans took them.[2]

Yet Hitler was circumspect; he had no reason for immediate conflict with the United States although he was mindful that Roosevelt had announced his support for Britain in the shape of material aid. Such a policy on the part of the Americans clearly infringed their position towards Germany under the terms of neutrality then in force and was thus in effect tantamount to a prosecution of war against Germany. Nevertheless, Hitler – his mind already turning to the East – resisted any temptation to provoke the United States at a time when he was engaged in the defeat and conquest of Britain, despite the fact that the USA was rendering Britain vital aid with which it could continue to fight. This was especially so in the North Sea and Atlantic where the toll on British merchant shipping was undeniably increasing. The amount of damaged tonnage far exceeded the resources available for repair, and with her yards becoming ever more congested, Britain was finding it difficult to carry her vital war supplies.

Churchill pushed Roosevelt for more support and eventually it came. The United States assisted by making available ship repair and docking facilities on her Eastern Seaboard as well as naval escort to British merchant vessels. The direct supply of materiel and other essential commodities was increased. Hitler knew this and was no doubt frustrated by it, but he lacked any tangible weapon or means with which to take action to arrest American intervention.[3]

Above In 1940, Grand Admiral Erich Raeder (right), the Commander in Chief of the Kriegsmarine (German Navy), supported by Göring (left) as commander of the Luftwaffe, urged Hitler to capture Iceland, the Azores and the Canaries before they could be utilised by the Americans. Raeder envisaged using the Atlantic islands as bases from which operations could be conducted against the United States. He also oversaw the policy of unrestricted U-boat warfare against Britain, but was fundamentally opposed to the later German invasion of Russia. Gradually his differences with Hitler resulted in his retirement in January 1943, when he was replaced by Karl Dönitz.

As the summer wore on, so the *Luftwaffe* battled with the RAF in the skies over the Channel and south-east England. On 20 July 1940, the day after Hitler had delivered his 'last appeal to reason' speech at the *Reichstag*, the newly promoted *Reichsmarschall* Hermann Göring held a press conference in Berlin for a group of American journalists. Questioned as to the possibility of a German threat to the USA, the *Reichsmarschall* scoffed: '*The United States cannot be invaded. The* Luftwaffe *has no planes of sufficient range for transatlantic operations. Even if you don't like us, give us some credit for common-sense and reason.*'[4]

Yet a short while later, on 10 August, the *Seekriegsleitung* (SKL – Naval War Staff) wrote to Göring advising him that in addition to naval bases, the envisaged 'Greater German Colonial Reich' in central Africa would need long-range aircraft capable of flying at least 6,000 km in order to offer air protection to German shipping and to carry priority supplies and fresh produce home.[5] Such a demand was echoed on 15 November when the minutes of a conference at the *Führer's* headquarters noted the need for a 'Bomber with a range of 6,000 km' which would be capable of operating from landing grounds built in the Azores. These islands had already been used by *Deutsche Lufthansa* before the war as a refuelling stop for transatlantic airliners. Indeed, by the winter of 1940, Hitler had recognised that the Azores – lying some 1,500 km west of Lisbon in the Atlantic – offered him the only viable base from which he could attack the United States with 'long-range Messerschmitts', thus forcing the Americans to set up their own air defence systems at the expense of helping the British.[6] On 24 August 1940, he had authorised Raeder to draw up a plan – code named '*Unternehmen Felix*' – to seize not only the Azores, but also Gibraltar and the Canary Islands. This plan assumed that Spain and Portugal would shortly enter the war on Germany's side. *Felix* did not materialise (nor did the Iberian countries join the Axis), but even as late as 29 October 1940, *Oberstleutnant* Sigismund *Freiherr* von Falkenstein, the *Luftwaffe* liaison officer with the Operations Staff of the OKW, issued a memorandum to unnamed senior *Luftwaffe* officers in which it was stated:

> '*The* Führer *is at present occupied with the question of the occupation of the Atlantic islands with a view to the prosecution of a war against America at a later date. Deliberations on this subject are being embarked upon... Essential conditions are at the present:*
>
> *(a) No other operational commitment; (b) Portuguese neutrality; (c) support of France and Spain.*
>
> *A brief assessment of the possibility of seizing and holding air bases and of the question of supply is needed from the* Luftwaffe...'[7]

The following month, Hitler was again discussing with the Staff of the OKW the need for a long-range bomber which could operate in support of the U-boat 'wolf packs' in the Atlantic and conduct operations against the United States. Worsening tensions with America and the increasing likelihood of air operations over the Atlantic, prompted the RLM to issue a requirement for an ultra-long-range bomber which could fly 12,000 km and undertake so-called '*Amerika Fall*' flights from Brest on the western coast of France to New York and back. Such missions would be necessary for the transport of express courier mail and vital raw materials for the war effort which were slowly, but surely, becoming scarce in Europe.[8]

Below The offices of the Messerschmitt A.G. Projektbüro (project department) at Augsburg. Much of the design and development work on the Me 264 was undertaken here.

At Augsburg, *Professor* Messerschmitt dusted off the files for his P 1061 project and summoned his project engineers *Dipl.-Ing.* Wolfgang Degel, *Dipl.-Ing.* Karl Seifert, *Dipl.-Ing.* Paul Konrad and *Dipl.-Ing.* Woldemar Voigt. These men, particularly Seifert and Konrad who had great experience in glider design having once run their own company at Rosenheim, would be involved in the detailed development work of the Me 264. From early 1941 to December 1943, Isolde Nacke – who would later marry the Messerschmitt test-pilot, Karl Baur – worked as an assistant to Karl Seifert and was involved in helping to produce several of the early engineering drawings of the Me 264; she recalls: 'Karl Seifert and Paul Konrad were half-brothers. *Dipl.-Ing.* Paul Konrad was the older of the two. During my entire life I never met any other two brothers who were as different as they were. Karl was a missionary; the humble, quiet, yet forward-looking scientist. Paul, however, was the total realist, always outspoken and a 'mechanic' with an eye for detailed design. On several occasions I witnessed Paul chiding his younger brother over his ideas, "*You are dreaming again, boy, come back down to earth!*" And Karl would answer quietly, "*It is possible – you will see...*" Yet as different as they were, each was brilliant in a technical way. Their characters complemented each other and they worked successfully as a team for many years.'[9]

Above The Daimler-Benz DB 603 engine with which Professor Messerschmitt planned to equip his 1940 proposal for an aerodynamically superior 'optimal aircraft' with a range of 20,000 km.

Messerschmitt outlined his proposal to his design team for using the Me 261 as a basis to further develop, in stages, an 'optimal aircraft' with a range of 20,000 km for use in both the long-distance civilian and military roles and powered by four of the new 12-cylinder DB 603 engines. With emphasis on 'optimal', the fuselage was to be as near aerodynamically perfect as possible. In the case of military usage, Messerschmitt saw a potential payload of two to five tons carried in a spacious internal bomb bay in the centre section of the fuselage with provision for smaller bombs to be carried on external racks under the wings. Defensive armament would be housed in retractable turrets which, when retracted and not required, were to lie absolutely flush with the fuselage as had been planned for the Bf 165.[10]

On 22 January 1941, the *Luftwaffe* General Staff called for comparison tests of long-range aircraft deemed suitable for supporting the U-boat campaign in the Atlantic. Available for such tests – in one way or another – were three types: the Focke-Wulf Fw 200 *Condor* airliner, adapted in only small numbers for anti-shipping work with KG 40, but struggling to deal with the rigours of operational flying; the twin-engined Heinkel He 177 bomber of which just three prototypes were available for testing at the *Erprobungsstellen*; and the huge Blohm & Voss BV 222 *Wiking* flying-boat of which only one example was available at this time. While the manufacturers of these aircraft worked to tender the best possible performance data, Willy Messerschmitt pressed his project engineers to complete their work on the proposed Messerschmitt Me 264 and also eagerly offered a set of performance and weight specifications for consideration. He even promised Udet that the first machine (the V 1, W.Nr. 0001) would commence flight tests as early as May 1942. Unlike his immediate superior, Milch, Udet must have been impressed by Messerschmitt's optimistic claims, for the Me 264 was accepted as the most favourable aircraft for long-range 'Atlantic operations' – remarkable considering that only a limited quantity of construction equipment and material had been agreed for the project.

Prudently however, the RLM erred on the side of caution and searched for an alternative tender from another manufacturer in case Messerschmitt's projected delivery dates proved flawed. One possibility came from Arado whose Ar E 470 project appeared feasible. Developed from the Ar E 340 twin-engined bomber project of 1940, the Ar E 470 was designed as a large transatlantic bomber with a gondola-type central fuselage and tapering twin tail booms connected by a horizontal tail unit and elevator. Arado proposed five variants, all heavily armed and powered by four or six 4,000 hp Daimler-Benz DB 613 24-cylinder engines.

Below In early 1941 the Luftwaffe General Staff called for comparison tests of long-range aircraft deemed suitable for supporting the U-boat campaign in the Atlantic. Willy Messerschmitt (centre in this photograph) pressed his project engineers to complete their work on the proposed Me 264 and eagerly offered a set of performance and weight specifications for consideration. He even promised Udet (to Messerschmitt's left) that the first prototype would commence flight tests as early as May 1942. Milch however (to Messerschmitt's right) was doubtful of the aircraft designer's promises and demanded revised specifications. Also seen here, behind Udet, is Fritz Hentzen, Messerschmitt's Production Manager and (far right) Hubert Bauer, Director of Operations.

Below By early 1941, the Focke-Wulf Fw 200 Condor airliner had been adapted in only small numbers for anti-shipping work with KG 40, but it was struggling to deal with the rigours of operational flying. Here, a pristine, early production Fw 200, W.Nr.0002, BS+AG, has been rolled out on a winter's day at the Focke-Wulf works in Bremen in early 1940. This machine was built as the V 11 and subsequently became the first B-1 variant. It was later converted to become the first C-1 and was taken on by 1./KG 40 on 19 February. Its subsequent fate is unknown.

Left The wind tunnel model of the Me 264, complete with gun turrets, on which Messerschmitt conducted tests in the spring of 1941.

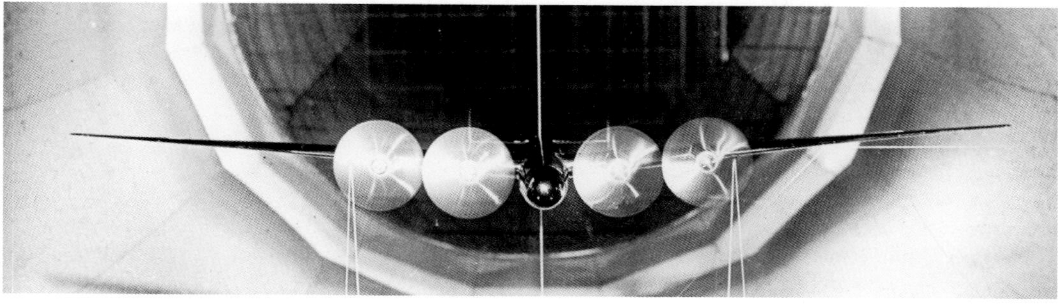

Left The wind tunnel model of the Me 264 fitted with free moving propeller strips. Note how the strips have been positioned so that they overlap slightly.

The aircraft would have had a wingspan of between 47.3 m and 68.5 m. A sixth variant, the 'F', was planned as a transport with a 15,000 km range and freight capacity of 39,000 kg in a container mounted below the wing centre section. With a maximum speed of 530 km/h and an operational ceiling of 11,000 m, variant 'E' would have been powered by four or six DB 613s in pairs with exhaust gas turbo-chargers and would have carried a 5,000 kg bomb load of the heaviest calibre bombs with a range of 14,900 km. A four-man crew was to have been accommodated in a pressurised cabin with remote-controlled defensive armament. On paper, such a machine would have been able to conduct bombing raids over the US East Coast, returning to a landing ground in France without refuelling, but the RLM scrapped the project without reason, although ensuing problems with the DB 613 engines may have put an end to further consideration.[11]

On 5 March 1941, the RLM, still under Udet's direction but probably under Hitler's unofficial prompting, issued Messerschmitt with a contract for 30 'four-engined aircraft with two-ton payload for harassment flights against the US'. Sufficient material was authorized for the immediate construction of the first six machines under the now official designation 'Me 264'.[12]

Sceptical of Messerschmitt's delivery estimates however, the RLM recalculated the range. In order to carry a 5,000 kg bomb load over 12,000 km, six single DB 603 engines would be needed, but even this revision would not permit non-stop flight from western France to the US East Coast. Simultaneously, Messerschmitt himself was at work on a six-engined variant of the proposed Me 264 under the designation, P 1075, an aircraft envisaged to undertake ultra-long-range reconnaissance and strike operations over Asia, Africa and America. However, due to under capacity at the Augsburg project office, design work on components such as the wings and tail unit was sub-contracted out to Fokker at Amsterdam.

Generalfeldmarschall Milch did not share the *Führer's* favourable opinion of Messerschmitt and was opposed to his company participating in the race to build a bomber intended for such operations. Milch believed that the job should be left to Dornier or Junkers – most probably the latter since he had overseen Junkers' nationalisation personally and thus could effect control over it. Nevertheless, while the RLM's engineers once again looked at the viability of in-flight refuelling of the aircraft 3,000 km from its base over the Atlantic, the Messerschmitt *Projektbüro* worked to improve the Me 264's airframe structure and enhance performance. This work, which continued into the spring of 1942, included such possibilities of towing one Me 264 to altitude using another such machine; in-flight refuelling by another Me 264 and the use of rocket-assisted take-off (RATO) units. The *Projektbüro* believed that with such enhancements, range with a 5,000 kg bomb load could be extended to 18,100 km, or up to 26,400 km with no load. Increased bomb loads and new weapons in the form of remote-controlled turrets fitted with either MG 131 or MG 151 *Zwilling* (twin or coupled two-gun) configurations were also assessed. The problem, however, was the continuing unavailability of the new DB 603 or 12-cylinder, 2,100 hp Jumo 213 engines which would have lent the aircraft the power it needed.

Despite this, the Messerschmitt design team doggedly completed preliminary plans for fuselage, wings and tailplane and a little later, in April 1941, extensive wind tunnel tests were carried out including a comparison with the twin-engined Me 261.[13]

On 4 April 1941, the eve of the German invasion of Yugoslavia and Greece, Hitler met with the Japanese Foreign Minister, Yosuke Matsuoka, in Berlin and assured him that '... Germany would wage a vigorous war against America with U-boats and the *Luftwaffe* against targets in the USA... the American Republic would have to be dealt with severely.'[14]

Hitler's view on the United States was hardening, but he needed Japan's co-operation to keep the Americans out of a European war until Germany was in a position to engage. 'The people in the US,' Joachim von Ribbentrop, Hitler's Foreign Minister had opined to the Japanese ambassador to Germany a few weeks earlier, 'were not willing to sacrifice their sons, and therefore were against any entry into the war. The American people felt instinctively that they were being drawn into war for no reason by Roosevelt and the Jewish wire-pullers. Therefore our policies with the US should be plain and firm...'

'America was confronted by three possibilities.' Hitler informed Matsuoka, 'She could arm herself, she could assist England, or she could wage war on another front. If she helped England she could not assist herself. If she abandoned England the latter would be destroyed and America would then find herself fighting the powers of the Three-Power Pact alone. In no case, however, would America wage war on another front.'

Hitler had grossly underestimated American capability. By 1941 the United States was comfortably in a position to supply not only its own markets with both industrial goods and agricultural produce, but it was able also to supply other nations, thus cementing its important alliances in the process. The vast Atlantic and Pacific Oceans protected America against expansionist threat from Imperial Japan or Nazi Germany. The United States was thus in a position to intervene in the far distant theatres of war in Asia or Europe as it wished.[15] Yet Hitler seemed only to have contempt for American military power; *Oberst* Gerhard Engel, Hitler's Army Adjutant, recorded the *Führer's* obsession with racial war in his diary on 24 March 1941: '*The Jews must be taught a lesson through terror attacks on the American cities.*'[16]

Albert Speer offers a further insight in his memoirs:

> '*The Americans had not played a very prominent part in the war of 1914-18, he* [Hitler] *thought, and moreover had not made any great sacrifices of blood. They would certainly not withstand a great trial by fire, for their fighting qualities were low. In general, no such thing as an American people existed as a unit; they were nothing but a mass of immigrants from many nations and races.*
>
> '*Fritz Wiedemann, who had once been regimental adjutant and superior to Hitler in his days as a courier... thought otherwise and kept urging Hitler to have talks with the Americans. Vexed by this offence against the unwritten law of the round table, Hitler finally sent him to San Francisco as German consul general. "Let him be cured of his notions there."*'[17]

Hitler returned to the notion of launching bombing raids against US East Coast and hinterland cities from the Azores during a conference with the SKL on 22 May 1941. He surprised those senior officers attending with his vision of sprawling runways on enormous airfields in the mid-Atlantic from which long-range heavy bombers could operate. In a reversal of his earlier position, Raeder attempted in vain to dissuade the *Führer* from his belief in such ideas, since the Navy did not have the capability to occupy the islands in the first place: 'The *Führer* is still in favour of occupying the Azores in order to be able to operate long-range bombers from there against the USA. The occasion for this may arise by autumn.'[18]

Two months later, buoyed by the startling success of Operation *Barbarossa*, the invasion of Soviet Russia, Hitler reiterated his policy towards the United States in a document to the Naval High Command:

> '*The military domination of Europe after the defeat of Russia will enable the strength of the Army to be considerably reduced in the near future. As far as the reduced strength of the Army will allow, its armoured units will be greatly increased.*
>
> '*Naval armament must be restricted to those measures which have a direct connection with the conduct of the war against England and, should the case arise, against America.*
>
> '*The main effort in armament will be shifted to the* Luftwaffe, *which must be greatly increased in strength.*'[19]

As Germany's armies advanced ever deeper into Russia, the demands on the resources of the *Luftwaffe* and the aircraft industry grew greater. Yet in the summer of 1941, the German High Command, believing in the prospect of another *Blitzkrieg* victory with the anticipated capture of Moscow, had not adequately foreseen the vast distances their soldiers, vehicles and aircraft would have to conquer. Thus, in Berlin, Udet busied himself with other matters such as the heavy bomber programme and how the *Luftwaffe* would be able to operate across the Atlantic. Udet eventually concluded that Germany lacked the bombers with which to inflict defeat on the USA via an air campaign alone. On the other hand, 'nuisance' or 'harassment' raids conducted by single or small formations of long-range aircraft against the East Coast might force the Americans to reinforce their coastal defences rather than despatching aid to Britain. Britain's ability to wage war against Germany would then be reduced. This was a hopelessly flawed idea originating from a man who remained committed *against* the heavy bomber. The American Government would simply have built the defences it needed without any need to reduce overseas aid. This would have made planned

Left Generaloberst Ernst Udet favoured Heinkel's He 177 as a suitable four-engined heavy bomber for the Luftwaffe in 1941, an aircraft, which featured two engines on each wing coupled together as one driving a single airscrew. This gave the aircraft the appearance of a twin-engined machine, and produced a low-drag aerodynamic configuration. However it suffered from enduring problems with its DB 606 power plant. Whilst the He 177 met the need for a heavy bomber, it fell short of being a long-range bomber. The machine seen here was the second pre-production aircraft, the He 177 A-02, coded DL+AQ. It was tested at Rechlin but its ultimate fate is unknown.

German raids even more challenging to mount. Any serious threat of air attack would have prompted the Americans to relocate vital industries under jeopardy further west, but in truth, US industry was very able to deal with an occasional 'pin-prick' nuisance attack.

Udet placed his hope in the Heinkel He 177 bomber but there was bad news in that particular direction. At an RLM production conference held the day before Hitler and Raeder had discussed using the Azores as a launch base from which to strike America, an officer from the *Luftwaffe's Versuchsstelle für Höhenflüge* (VfH – Experimental Testing Station for High-Altitude Flight), one of the units involved in testing the aircraft, reported that in the unit's opinion the He 177 in its twin-tandem DB 606 engine configuration was troublesome, unreliable and unsuitable for operations. Four independently-mounted DB 603 or BMW 801 engines were deemed preferable, but this would entail increasing the wing area – and thus weight – by another 18 square metres. Furthermore, the anticipated range of between 4,500-5,000 km already ruled out sorties any further than the mid-Atlantic. In the He 177, the *Luftwaffe* had a *heavy* bomber, but not a *long-range* strike aircraft,

Plans were put forward in August to equip the He 177 – an aircraft now considered to be 'of the greatest importance' – for in-flight refuelling. It was calculated that if an He 177 refuelled in the air it would extend its range to 9,500 km, enough to patrol the distant Atlantic skies for American bombers being ferried across to Britain, but still not enough to reach and attack the American coast.

In essence, the *Luftwaffe* had suffered from the start in not having sufficient plans in place for operations beyond tactical support of the Army and what finally persuaded it to re-examine the deployment of a heavy bomber was the failure of the Ju 88 in the air war over Britain in 1940. The nimble Junkers had failed to make an appreciable impact either on the British civilian population or in terms of military results. However, despite these initial failures, ensuing successes in the Balkans – victories reminiscent of the earlier campaigns – and initial success in Russia, served to return *Luftwaffe* thinking – driven by Udet and Jeschonnek, the *Luftwaffe* Chief of General Staff – to limited strategic operations with emphasis on more immediate tactical requirements.[20]

Meanwhile, with Milch's reluctant agreement, Messerschmitt continued to work quietly on the Me 264 at Neu-Offing near Ulm. In August 1941, with typical optimism, he surprised all those involved by announcing that the first prototype aircraft would be completed and available for flight-testing by early summer 1942. This surprised Udet more than others; the calculations lying on his desk indicated that the aircraft would only be able to reach the US coastline and that targets inland were

Below Seen at Augsburg for a flight demonstration in 1941, from left: Professor Willy Messerschmitt (in civilian hat and overcoat), Reichsmarschall Hermann Göring (centre), Fritz Hentzen (Messerschmitt's Production Manager in hat and overcoat), Generaloberst Ernst Udet (in leather greatcoat). By late 1941, Udet realised that Messerschmitt had failed to attain his own targets of improved aerodynamic performance and range for the Me 264. In mid-November however, the Generalluftzeugmeister committed suicide as a result of increasing tensions with Göring.

By August 1941 construction work had reached an advanced stage on the Messerschmitt Me 264 V 1 at Augsburg. This sequence of photographs shows the assembly of the cockpit section and forward fuselage.

Right and below Two views of the partially completed cockpit housing in its jig. The somewhat crudely cut hole in the close-up is the initial 'cut' for the nosewheel bay.

Above and above right Two views of the cockpit section sitting in its wooden cradle prior to assembly to the main fuselage. Note in the photograph to right the prominent nosewheel leg housing protruding into the glazed section.

Below The extensive glazing of the Me 264's cockpit which was intended to offer maximum visibility to the crew, can be seen to advantage in this view of the nose section of the aircraft. Note the large nosewheel housing.

Above The nosewheel, struts, leg and hydraulics system have been assembled to the main fuselage. This assembly was to prove one of the weak points of the Me 264's design and would be problematic during flight-testing between 1942-44. Note the pilot's control column and the back of the instrument panel immediately behind the nosewheel hydraulics.

Above A view across the forward fuselage from the starboard side, showing the nosewheel housing and extended wheel. The port-side outboard Jumo 211 engine is also visible.

Right A detail view of the nosewheel and leg of the Me 264 V 1. The wheel had a diameter of 935 mm and was 345 mm wide. The wheel was anchored by forked suspension struts, and retracted by turning rearwards into a bay beneath the cockpit. Later, during flight testing, it would suffer from vibration and shimmying problems.

Above and above left Scale is given by these views of a technician consulting an assembly plan inside the cockpit area. In these photographs, although the nosewheel housing has been fitted, the control and instrument panel seen in the photograph at top left on the opposite page, appear to have been removed.

Right and far right Two further views of the nosewheel assembly also showing the installation of the control column and the mass of wiring spilling from the cockpit's central instrumentation console. The sign hanging from the top of the fuselage states: 'Achtung! Nicht auf das Gestänge steigen' – 'Attention! Do not climb on the struts'.

Above Following Ernst Udet's suicide on 17 November 1941, Generalfeldmarschall Erhard Milch assumed the post of Generalluftzeugmeister. Ambitious and known to be quite ruthless and insensitive, his animosity towards Messerschmitt would serve to cast a long shadow over the whole Me 264 development programme. From the mid-war period, Milch attempted to persuade Göring to reduce bomber production in favour of fighters, but Göring often showed little interest in his reports. The fact that Milch's mother was Jewish did not stop his meteoric rise to the very top of German aviation, and ultimately, along with Albert Speer, in control of transportation throughout the Third Reich.

beyond range. Messerschmitt had thus failed to attain his own targets of improved aerodynamic performance and range.

Within three months however, Udet, anguished and depressed, was dead, having shot himself on 17 November as a result of the stress and pressures of his position, a feud with the dominant Göring and the shortcomings of the *Luftwaffe's* failures on the Eastern Front. The ambitious Erhard Milch then took over the post of *Generalluftzeugmeister*.

On 7 December 1941, nearly 400 Japanese aircraft attacked American warships at Pearl Harbor and the United States entered the Second World War. 'The turning point!' Hitler had exclaimed to Martin Bormann and other members of his inner circle.[21]

To many of those in the *Führer's* court, it seemed that America's entry into the war was driven by greed, self-confidence and pride. Such was his disdain that Hitler was heard to comment: '...one Beethoven symphony contains more culture than America has produced in her whole history!' [22] From certain information he was receiving from his diplomats in Washington, Hitler had evidence that Roosevelt was reluctant to be drawn into a war. Yet hostilities between Japan and the USA would serve a purpose – to distract the Americans with war in the Pacific, at least for the coming year, thus allowing Germany to conclude its war in Russia.[23]

Four days later, on 11 December, Germany declared war on the United States; the United States was still neutral in Europe.

That month at Augsburg, mindful of his 'summer 1942' promise, Messerschmitt rushed out a specification brochure on the Me 264 entitled '*Leistungsangaben, Gewichte, Ansichtzeichnungen Me 264*' which contained detailed artists' impressions of the planned long-range bomber, as well as variants intended as 'heavy bomber' and 'long-range reconnaissance'. The bound, landscape booklet – the first known 'public' glimpse of the Me 264 – contained parts lists, weight distribution data, curves on engine performance, speeds and take-off performance, range data, and rates of climb and ceiling.[24] Messerschmitt A.G. stipulated that certain fundamental design and equipment requirements would be necessary, irrespective of whether the aircraft was to be used for military purposes or not, such as the most up-to-date navigation and radio equipment, the fitting of auxiliary drop fuel tanks, maximum accommodation area, a pressurised cabin and heating. For military requirements, the design would need to include hand-operated and remotely-controlled defensive armament, armour, armoured/protected fuel tanks and interchangeable equipment to allow the aircraft to be able to test special equipment or to undertake different types of operations with minimum alterations.

Six basic military operational variants were proposed:

a) Me 264 H 3 *Langstreckenbomber* (long-range bomber): A long-range bomber fitted with four DB 603 H engines, with a take-off weight of 50 tons, bomb load up to 8,400 kg and a range of 15,600 km without payload and 11,300 km with bomb load

b) Me 264 H 3 *Langstreckenbomber* (long-range bomber): A long-range bomber fitted with four Jumo 213 engines, with a take-off weight of 50 tons, bomb load up to 8,400 kg, and a range of 15,600 km without payload and 11,600 km with bomb load

c) Me 264 H 3 *Schwerer Bomber* (heavy bomber): A heavy bomber fitted with four DB 603 H engines, with a take-off weight of 50 tons, a main bomb load and an auxiliary bomb load carried on the wings equating to a total payload of 14,000 kg. Range 11,500 km without payload and 7,900 km with bomb load

d) Me 264 H 3 *Schwerer Bomber* (heavy bomber): A heavy bomber fitted with four Jumo 213 engines, with a take-off weight of 50 tons, a main bomb load and an auxiliary bomb load carried on the wings equating to a total payload of 14,000 kg. Range 11,900 km without payload and 8,400 km with bomb load

e) Me 264 H 2/H 3 *Fernaufklärer* (long-range reconnaissance): A reconnaissance machine, patrol aircraft or courier aircraft fitted with four Jumo 213 engines (H 2) or DB 603 engines (H 3) with a 57 ton take-off weight. Projected range 20,800 km. Flight duration approximately 60 hours

f) A long-range transport aircraft.

Messerschmitt had designed the aircraft with a wingspan of 43.1 metres, an overall fuselage length of 20.55 metres, a height of 4.28 metres and a total wing area of 125 square metres. Though the standard power units for these variants were envisaged as DB 603 and/or Jumo 213 engines, it would be possible to use alternative types such as the 14-cylinder, 2,000 hp high-altitude BMW 801 radial and 24-cylinder, 2,500 hp Jumo 222, as well as high-altitude engines such as the DB 614 or Jumo 223, with which the maximum attainable ceiling would be 13-14 km. Extendable engine coolers were to be built in at an angle to the power units and protected by their position between the engines and the armoured wing fuel tanks. In such a position the cooler would be protected to some extent from enemy fire. It was envisaged that each engine would have a lubricating oil tank which would draw oil from the main tank in the heated fuselage.

The aircraft was to be built as an all-metal, semi-high-wing monoplane with retractable landing gear and nosewheel. The fuselage was of monocoque form, divided into four equal internal parts. The section from the rear of the cockpit to the tail assembly was built as one unit, with the area around the bomb-bay and a part of the navigator's area, being cylindrical. The fuselage cross-section was circular. Adaptable construction techniques included removable and interchangeable cockpit units. The tail assembly was copied from that used on the Me 261. The aircraft could also be built easily in a civilian role with the

Above In August 1941, Willy Messerschmitt announced to Milch and Udet that the Me 264 would be available for flight-testing in the summer of 1942. To overcome doubts and scepticism about the likelihood of such a date, in December 1941, Messerschmitt A.G. at Augsburg produced a detailed specification brochure on the aircraft which featured artists' impressions and data tables on the planned variants. This is the cover of that brochure featuring a stylized sketch of the Me 264, complete with remotely-controlled rearward firing 20 mm MG 151 machine guns mounted on the trailing edge of the wing immediately behind the inboard engines, plus at least one upper fuselage turret equipped with an MG 131 Zwilling with what appears to be another just aft of the cockpit, probably above the navigator's position. The engines depicted appear to be based loosely on Daimler-Benz in-line engines, possibly the DB 603.

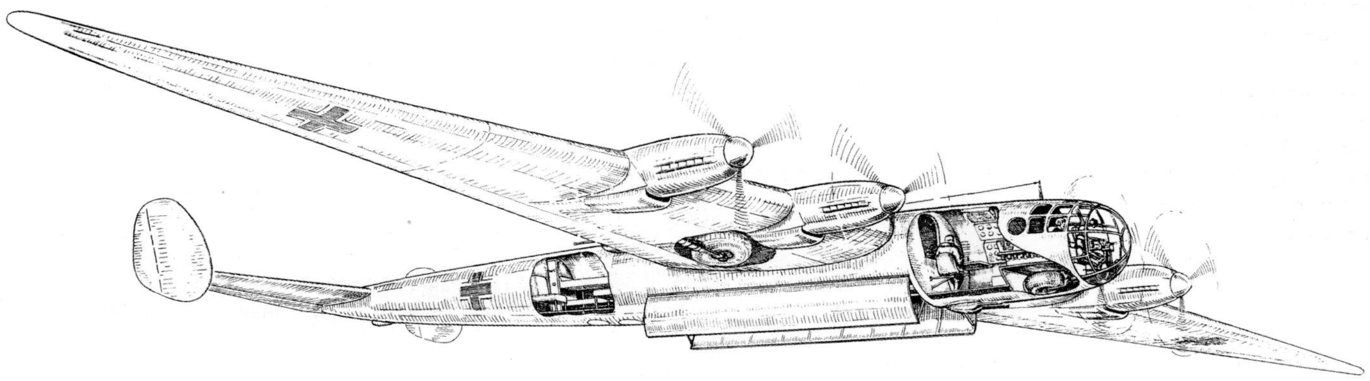

Above A cut-away sketch of the Me 264 as seen in Messerschmitt's 1941 promotional brochure for the aircraft. Note the 'all-round' visibility cockpit glazing and the spacious navigator's compartment behind the cockpit. Aft of the bomb bay, the doors of which are open, is the crew rest area with entrance to a toilet. Note also the under-fuselage remote-controlled 13 mm MG 131 Zwilling (twin) turret immediately below the navigator's position. The artist has also shown the nose and mainwheels retracted, the nosewheel lying flat in its bay, having turned rearward as it retracts. What is missing from this sketch is the large nosewheel housing which protruded into the cockpit on the eventual Me 264 V 1.

Below A dynamic view of the Me 264 taken from the Messerschmitt December 1941 specification brochure, depicting the aircraft as it might appear taxiing towards take-off. In this view the tricycle undercarriage is shown to advantage. The nosewheel, which would measure 935 mm x 345 mm, was held by forked suspension struts, while the four mainwheels each measured 1,550 mm x 575 mm. The outer mainwheels, included to give the aircraft sufficient support during take-off, were jettisonable.

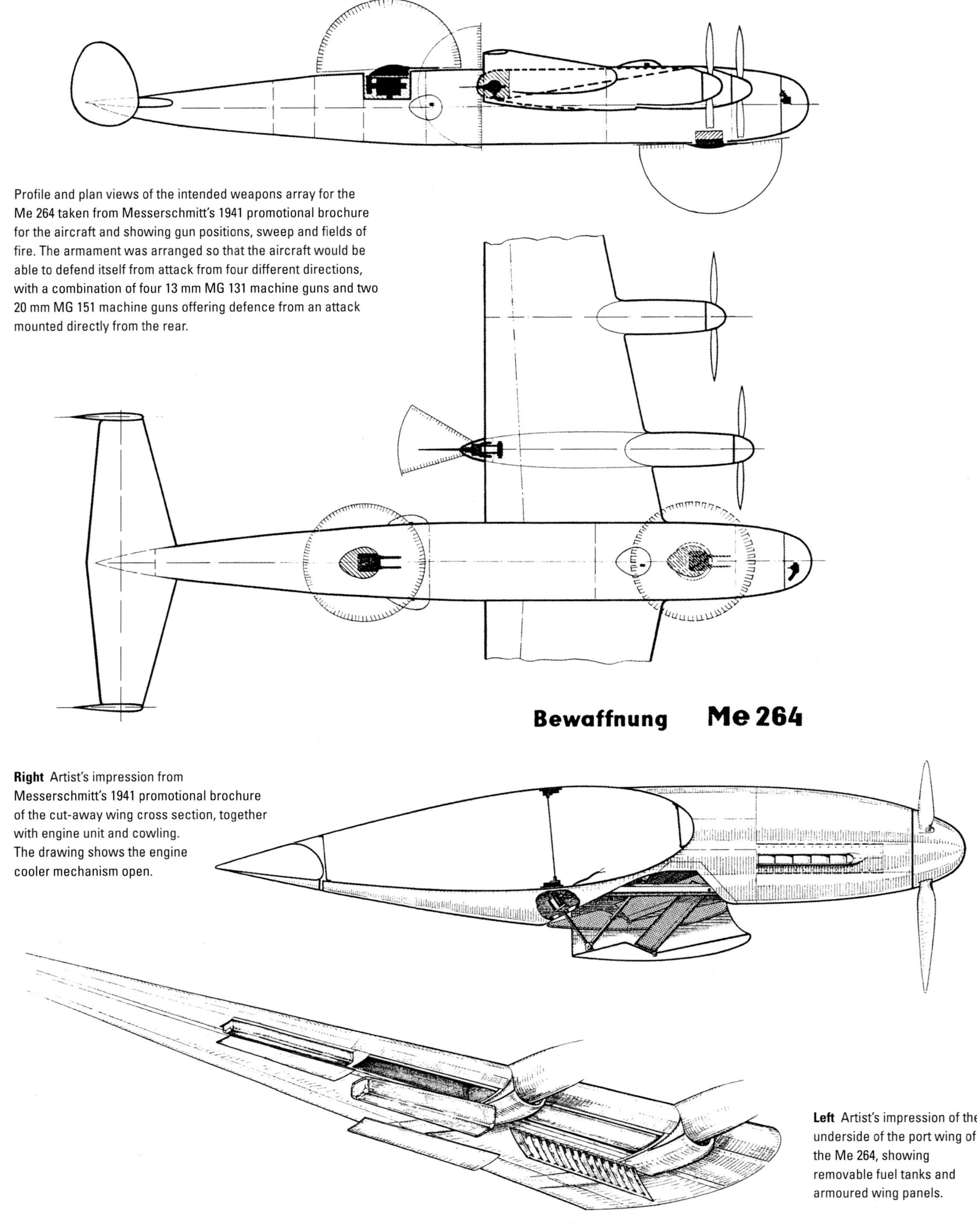

Profile and plan views of the intended weapons array for the Me 264 taken from Messerschmitt's 1941 promotional brochure for the aircraft and showing gun positions, sweep and fields of fire. The armament was arranged so that the aircraft would be able to defend itself from attack from four different directions, with a combination of four 13 mm MG 131 machine guns and two 20 mm MG 151 machine guns offering defence from an attack mounted directly from the rear.

Right Artist's impression from Messerschmitt's 1941 promotional brochure of the cut-away wing cross section, together with engine unit and cowling. The drawing shows the engine cooler mechanism open.

Left Artist's impression of the underside of the port wing of the Me 264, showing removable fuel tanks and armoured wing panels.

inclusion of additional navigation equipment, auxiliary fuel tanks, and extra accommodation as required, or as a military machine with pressure cabin, armour and navigation equipment as required, armoured main fuel tanks, auxiliary fuel tanks, and provision to carry external ordnance such as aerial torpedoes and bombs.

The wings, which were similar in profile to those of the Me 261, were trapezoid in shape and strength came from one main spar and two support spars built into the fuselage. Each wing would hold nine fuel tanks, of which six were armoured. Fuel was pumped internally by a pressure pump. Fuel from the non-armoured tanks was used first by pressure pumping from a CO^2 tank in the rear fuselage. Each engine would be fitted with a fire extinguisher.

The aircraft should have been able to carry sufficient fuel in its armoured tanks to allow it to fly for 60 to 100 per cent of its range after take-off with a bomb load, although range could be increased with the inclusion of external and/or emergency tanks.

The undercarriage comprised four 1550 mm x 575 mm mainwheels in pairs, of which one wheel from each pair was jettisonable after take-off. The wheels turned on their axis as they retracted electro-hydraulically. A 935 mm x 345 mm nosewheel, held by forked suspension struts, retracted turning rearwards into a bay beneath the cockpit.

Entrance to the aircraft for the crew was via an angled access hatch beneath the crew accommodation area.

The roomy, comfortable cockpit offered good vision above, below and to 60 degrees either side for both pilot and co-pilot and contained all flying instruments, controls and steering systems together with the emergency bomb release system. Control of the aircraft could be taken from either the pilot's or co-pilot's positions, whose armoured seats were adjustable, forward and back, and by height.

Adjacent to the cockpit was the navigator's position with a navigation console and table with equipment for astro-navigation, as well as the radio equipment. An emergency exit was installed above the navigator's position for use in the event of a belly landing. From this point, a gangway led to the crew's heated bunk and rest area aft of the bomb bay, where there was also camera equipment and a toilet. Sleeping accommodation was provided by four bunks.

To enable repair and maintenance, there were large open access hatches built into the walls forward and aft of the bomb bay and throughout the aircraft.From the time of this December 1941 booklet, it is clear that Messerschmitt always saw the Me 264 being equipped with remotely-controlled armament and gun positions on the forward and central upper fuselage, below the cockpit and on the trailing edge of the wing immediately behind the in-board engines. The proposed armament array would allow the aircraft to defend itself from attack from four different directions, while a combination of four MG 131s and two MG 151s would offer defence from an attack mounted directly from the rear.

By the turn of the year, *Generalfeldmarschall* Milch placed manufacturing priority on the production of more urgently needed fighters, tactical bombers and transports to service the escalating demands of Germany's multi-front war. A long-range bomber was no longer considered a priority. With his dislike of Messerschmitt publicly unabated, Milch ordered the number of Me 264 prototypes to be cut and the earlier approved total of 24 series aircraft to be stopped. It has been suggested, however, that in terminating the agreement with Messerschmitt, Milch was hoping to be able to pass whatever development work remained on a long-range bomber to either Kurt Tank of Focke-Wulf who had submitted his six-engined Ta 400 proposal or to Ernst Heinkel for an uprated version of the He 177.[25] Indeed, Milch was out for blood; at the end of 1941, the *Luftwaffe* had pronounced Messerschmitt's Me 210 multi-purpose *Zerstörer*/ dive-bomber/ reconnaissance and ground-attack aircraft totally unsuitable for operational use although the *Generalluftzeugmeister's* office remained reluctant to order a termination of the aircraft's production because there was no aircraft in prospect to replace it. Matters were brought to a head when Milch ordered a halt to Me 210 production in favour of the more reliable Bf 110 pending a solution of the aircraft's many unsolved problems.

With this the case, it is somewhat curious that at exactly this time Milch's department at the RLM issued an official *Flugzeugtypenblatt* – Aircraft Type Sheet – for the Me 264. This drawing showed the aircraft fitted with remotely-operated rearward-firing MG 151s installed on the wing trailing edge directly behind the inboard engine nacelles. Two twin MG 131 *Zwilling* gun sets were built into turrets on top of the forward and rear fuselage.

The RLM then received revised performance data, projecting a range of 13,000 km based on the DB 603 engine. The bomb load was given as 3,000 kg with the weight of defensive armament at 2,000 kg. Six BMW 801 engines were proposed as an alternative power source forcing a reduced payload of 2,000 kg, but without reduction in armament, the range would decrease to 12,000 km. However if Jumo 211 in-line engines were used, even without a bomb load, range would struggle to reach 11,000 km.

The reality was that even if Milch had looked on the project more favourably, the engines were not available in the short-term to progress with these plans anyway. The best estimate for delivery of the Me 264 was 1946/47. Again, the RLM effectively abandoned the Me 264 as a serious proposal to look for an aircraft which could become available – possibly the Ju 290.[26]

Meanwhile, by the beginning of February 1942, work on the Me 264 prototypes was proceeding very slowly, mainly due to a lack of production capacity at the Augsburg works. It had been a difficult month for Messerschmitt. The imbalance between the purchase costs of new materials, half-finished Me 210 components and advance payments from the RLM amounted to 25 million RM, while his company's monthly overheads were running at 16 million RM. After 17 Me 210s had been lost in one week undergoing tests and faced with the very real prospect of bankruptcy for a second time, Willy

continued on page 39

Above A mock-up of the Me 264 V 1's cockpit with plywood instrument panels and centre console. This was to assist in achieving optimum design and space. The pilot's and co-pilot's seats are also temporary, with the pilot's seat being cushioned and leather-backed. Note how it has been badly scratched at some stage and also the seat position adjustment lever at the base of the seat to the right. Also missing from this mock-up is the large nosewheel housing. The cockpit glazing frame is not the one finally used in the V 1. The control column steering grip is very rudimentary and is not the final equipment to be used in the aircraft. The only dial installed into the central instrument console is a compass, and the throttle levers are, again, very provisional. A 13 mm MG 131 machine gun with Revi 16B reflector sight has been installed on a mount in the nose directly ahead of the co-pilot's position and an ammunition belt hangs from a feed chute above. There is a cushion immediately in front of the gun. Compare this photograph with the one of the virtually completed cockpit on page 70.

Below View of the timber mock-up of the co-pilot's instrument console, before the installation of the co-pilot's seat on its runners. Considerable natural light would have been available in the V 1 to aid control when flying during hours of daylight due to the proximity of the instruments to the cockpit's window panels.

Below An HD 151 electro-hydraulically-powered gun turret, containing a 20 mm MG 151 cannon as proposed for the Me 264 V 1.

The proposed Me 264 *Langstreckenbomber* H 3 was to be a long-range bomber with the following specifications:

Engine:	Daimler-Benz DB 603 (fuelled with C3 100 Octane)
Propeller Diameter:	3.9 m
Airframe	
Span:	43.1 m
Length:	20.55 m
Height:	4.28 m
Wing area:	125 m^2
Armament	
B-Stand:	2 x MG 131 (1,000 rnds each)
C-Stand:	2 x MG 151 (1,000 rnds each)
D-Stand (x2):	1 x MG 151 (2 x 1,000-1,500 rnds each)
Weights	
Structural weight:	18,089 kg
Crew – six men:	540 kg
Crew supplies:	100 kg
External racks and tanks:	200 kg
Armour for two pilots:	100 kg
Bomb gear:	250 kg
Weapons:	1,000 kg
Bombs:	8,400 kg
Fuel for armoured flight:	14,300 kg
Fuel for unarmoured flight:	5,421 kg
Fuel – auxiliary tanks:	0 kg
Oil:	1,000 kg
Weight on ground:	20,079 kg
Take-off weight:	50,000 kg

Range with bomb load in fuselage

6 x PC 1400 (8,400 kg)	11,300 km
2 x SC 2500 (5,000 kg)	13,200 km
4 x SC 1000 (4,000 kg)	13,800 km
4 x LMB III (4,000 kg)	13,800 km
2 x SC 1800 (3,600 kg)	14,000 km
6 x SC 500 (3,000 kg)	14,400 km
11 x SC 250 (2,750 kg)	14,500 km
2 x LMF (2,000 kg)	15,000 km

(To carry 1,000 kg additional armour or weapons, a decrease in bomb load by approximately 1,800 kg would maintain ranges as above)

Take-off distance without auxiliary thrust:	2,050 m
With 4 x 1,000 kg thrust units:	1,320 m
Take-off performance rating:	5 m per second
Climb rating:	3.5 m per second
Normal cruise rating:	2 m per second
Maximum speed at 6,000 m on emergency performance:	610 km/h
Service ceiling after half range at optimum cruising speed:	9,000 m
Flight duration with 8,400 kg bomb load:	30.6 hrs
Landing speed:	134 km/h

Wing loading

Landing:	160 kg/m^2
Take-off:	400 kg/m^2

The proposed Me 264 *Schwerer Bomber* H 3 was to be a heavy bomber with the following specifications:

Engine:	Daimler-Benz DB 603 (fuelled with C3)
Propeller Diameter:	3.9 m
Airframe	
Span:	43.1 m
Length:	20.55 m
Height:	4.28 m
Wing area:	125 m2
Armament	
B-Stand:	2 x MG 131 (1,000 rnds each)
C-Stand:	2 x MG 151 (1,000 rnds each)
D-Stand (x2):	1 x MG 151 (2 x 1,000-1,500 rnds each)
Weights	
Structural weight:	18,089 kg
Crew – six men:	540 kg
Crew supplies:	100 kg
Bomb gear:	500 kg
Armour for two pilots:	100 kg
Weapons:	1,000 kg
Ammunition:	1,000 kg
Bombs:	14,000 kg
Fuel for armoured flight:	14,071 kg
Fuel for unarmoured flight:	0 kg
Oil:	800 kg
Weight on ground:	20,329 kg
Take-off weight:	50,000 kg

Range with bomb load in fuselage

10 x PC 1400 (14,000 kg)	8,150 km
2 x SC 2500 (13,200 kg) + 4 x SC 1800	8,600 km
4 x SC 1800 (10,800 kg)	9,900 km
8 x SC 1000 (8,000 kg)	11,500 km
8 x LMB III (8,000 kg)	11,500 km
2x SC 2500 + 4 X LT 1500	11,500 km

(To carry 1,000 kg additional armour or weapons, a decrease in bomb load by approximately 1,800 kg would maintain ranges as above)

Take-off distance without auxiliary thrust:	2,050 m
With 4 x 1,000 kg thrust units:	1,320 m
Take-off performance rating:	5 m per second
Climb rating:	3.5 m per second
Normal cruise rating:	2 m per second
Service ceiling after half range at optimum cruising speed:	610 km/h
Service ceiling at normal rating:	9,500 m
Flight duration with 8,400 kg bomb load:	30.8 hrs
Landing speed:	136 km/h

Wing loading

Landing:	160 kg/m2
Take-off:	400 kg/m2

Above A pair of SC 250 bombs with their fin sections removed. The Langstreckenbomber – long-range bomber – version of the Me 264 could carry 11 such bombs and maintain a range of 14,500 km. Such bombs were used against railway targets, bridges and underground installations up to 8 m depth.

Above An SC 1800 bomb sits on its trolley. Two of these bombs could be carried in the Me 264 Langstreckenbomber which could achieve a 14,000 km range with such a load, while four could be carried by the Schwerer Bomber variant, allowing a range of 9,900 km. Filled with 1,000 kg of Trialen explosive, such bombs were intended for use against groups of buildings and large merchant ships.

This page The large bomb bay of the Me 264 meant that it could accommodate a formidable selection of 'SC' class 'Minenbomben' or normal high-explosive demolition bombs. These were thin-walled bombs containing approximately 50 per cent explosives. They were used mainly for explosive effect against buildings.

Left A Luftwaffe armourer sits on a SC 1000-A bomb. The Me 264 long-range bomber was designed to carry four such bombs, allowing a range of 13,800 km, while the Me 264 'Schwerer Bomber' or 'heavy' bomber would be able to accommodate eight, for a range of 11,500 km. The bomb was intended for deployment against maritime targets and against large unprotected land targets.

Left The presence of a Luftwaffe airman in this photograph indicates the massive size of the SC 2500 which was designed for use against factories and complexes of buildings as well as large merchant ships. Two such bombs could be accommodated in the bomb bay of the Me 264 Schwerer Bomber, which would also be able to carry four LT 1500 torpedoes on wing racks in such a configuration and achieve a range of 11,500 km. The Me 264 Langstreckenbomber could carry two SC 2500s, achieving a range of 13,200 km.

Bomb bay and external ordnance configurations were planned with the following options:

Type	Fuselage Bay	External Wing Racks	Total Weight
High Explosive	11 x SC 250	8 x SC 250	4,750 kg
	6 x SC 500 plus 3 x SC 250	4 x SC 500	5,750 kg
	4 x SC 1000	4 x SC 1000	8.000 kg
	2 x SC 1800 plus 3 x SC 250	4 x SC 1800	11,550 kg
	2 x SC 2500 plus 3 x SC 250	Nil	5,750 kg
Air Mines	4 x BM 1000	4 x BM 1000	8,000 kg
Armour-Piercing	6 x PC 1000 plus 3 x SC 250	4 x PC 1000	10,750 kg
	6 x PC 1400	4 x PC 1400	14,000 kg
Mines	2 x LMF plus 3 x SC 250	4 x LMF	6,750 kg
	4 x LMB III	4 x LMB III	8,000 kg
Torpedoes	Nil	4 x LT 1500	6,000 kg
Radio/signal beacons	Nil	4 x Gobi	4,400 kg

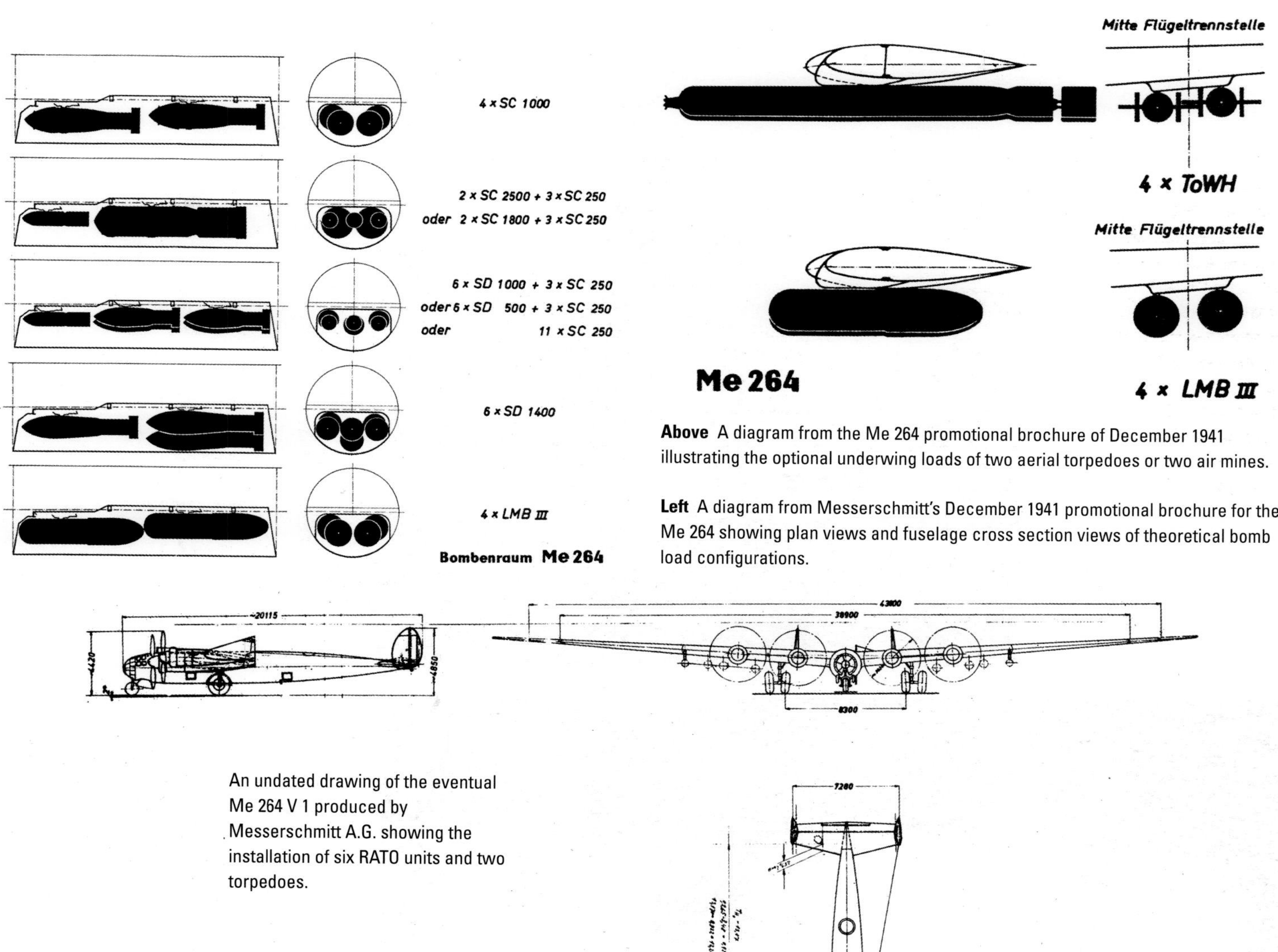

Above A diagram from the Me 264 promotional brochure of December 1941 illustrating the optional underwing loads of two aerial torpedoes or two air mines.

Left A diagram from Messerschmitt's December 1941 promotional brochure for the Me 264 showing plan views and fuselage cross section views of theoretical bomb load configurations.

An undated drawing of the eventual Me 264 V 1 produced by Messerschmitt A.G. showing the installation of six RATO units and two torpedoes.

The proposed Me 264 *Fernaufklärer* H 2 was to be a reconnaissance machine with the following specifications:

Engine: Jumo 213 A/2 (fuelled with C3)

Propeller Diameter:	3.9 m

Airframe

Span:	43.1 m
Length:	20.55 m
Height:	4.28 m
Wing area:	125 m2

Weapons
(unarmed)

Weights

Structural weight:	18,089 kg
Crew – six men:	540 kg
Crew supplies:	100 kg
Camera equipment:	140 kg
External containers:	300 kg
Fuel for armoured flight:	14,300 kg
Fuel for unarmoured flight:	6,200 kg
Fuel in fuselage:	5,000 kg
Fuel – auxiliary tanks:	9,951 kg
Oil:	2,380 kg
Weight on ground:	18,869 kg
Take-off weight:	57,000 kg

Range: 20,800 km

Take-off distance without auxiliary thrust:	2,700 m
With 4 x 1,000 kg thrust units:	1,900 m
Climb rating:	3.7 m per second
Normal cruise rating:	2.3 m per second
Maximum speed at 6,000 m on operational performance:	610 km/h
Service ceiling after half range at optimum cruising speed:	8,500 m
Flight duration:	60 hrs
Average cruising speed:	350 km/h

Wing loading

Landing:	150 kg/m2
Take-off:	455 kg/m2

The proposed Me 264 *Fernaufklärer* H 3 was to be a reconnaissance machine with the following specifications:

Engine: Daimler-Benz DB 603 (fuelled with C3)

Propeller Diameter:	3.9 m

Airframe

Span:	43.1 m
Length:	20.55 m
Height:	4.28 m
Wing area:	125 m2

Weapons
(unarmed)

Weights

Structural weight:	18,089 kg
Crew – six men:	540 kg
Crew supplies:	100 kg
Camera equipment:	140 kg
External containers:	100 kg
Fuel for armoured flight:	14,300 kg
Fuel for unarmoured flight:	6,200 kg
Fuel in fuselage:	5,000 kg
Fuel – auxiliary tanks:	3,551 kg
Oil:	1,980 kg
Weight on ground:	18,869 kg
Take-off weight:	57,000 kg

Range:	17,560 km

Take-off distance without auxiliary thrust:	2,050 m
With 4 x 1,000 kg thrust units:	1,320 m
Take-off rating:	5.0 m per second
Climb rating:	3.5 m per second
Normal cruise rating:	2.0 m per second
Maximum speed at 6,000 m on operational performance:	610 km/h
Service ceiling after half range at optimum cruising speed:	8,600 m
Flight duration:	50 hrs
Average cruising speed:	350 km/h

Wing loading

Landing:	150 kg/m2
Take-off:	400 kg/m2

Above With the delivery date for the first Me 264 prototypes anticipated to be four, or even five years away, in early 1942, the RLM began to look for alternatives. One contender was the Junkers Ju 290, a four-engined transport to be powered by BMW 801 engines with a range of some 6,500 km. The Ju 290 V 1, W.Nr. 90 0007, BD+TX, seen here in its sleek profile view, had been created from the Ju 90 V 11 by increasing its wingspan and converting its circular fuselage windows to rectangular ones. The Ju 290 V 1 commenced flight tests in August 1942, but was later lost when it crashed on take-off at Pitomnik airfield on 13 January 1943 during the attempt to supply the encircled Sixth Army at Stalingrad. The aircraft had been assigned to KGr.z.b.V. 200 and was piloted by Flugkapitän Walter Hönig. Later variants of the aircraft conducted long-range maritime reconnaissance over the Atlantic with Fernaufklärungsgruppe 5.

Messerschmitt finally admitted to Göring and Milch that his company was struggling. On 25 March, Messerschmitt made a personal appeal over the Augsburg factory loudspeaker system exhorting the workforce to ignore rumours circulating about the uncertain future of the Me 210. The decision was then made to transfer construction of the Me 264 to Dornier, but bottlenecks in that company's production necessitated a further move to the Weser *Flugzeugbau*. However, to take on the Me 264, this smaller company required a new production line to be established in southern Germany and the idea came to nothing. Production of the Me 264 subsequently fell even further behind schedule.

On 15 February, Milch called *Flugbaumeister* Walter Friebel, head of GL/C-E2 and the engineer responsible for aircraft development, to his office and, with *General der Flieger* Hans Jeschonnek present, raised the question of the so-called '*Aufgabe Amerika*' (the 'Amerika Task'). Friebel was forthright; in his view, neither the Me 264 nor any Heinkel or Focke-Wulf project would have sufficient range to reach the East Coast of America. In Friebel's view such missions could only be achieved if an in-flight refuelling method was available. Jeschonnek was sceptical, although in-flight refuelling tests involving an Fw 58 and a Ju 90 had proved successful.[27]

It was not only in the offices of the RLM in Berlin that plans were under discussion to strike the United States. In Italy during the spring of 1942, Nicolo Lana, the very experienced civilian chief test pilot of the Piaggio aircraft company submitted a proposal to the Italian Air Ministry whereby he and a flight engineer would bomb central New York with a single 1,000 kg bomb while flying the three-engined Piaggio P.23R, MM.285, an aircraft which had broken various pre-war speed and distance records. On 30 December 1938, this aircraft, powered by Piaggio P.XI RC40 engines and carrying a 5,000 kg bomb load, had captured the world speed record by flying at 404 km/h for distances over 1,000 and 2,000 km. The aircraft was then moved to the Guidonia Test Centre where it remained unused for several months.

Left The Piaggio P. 23R seen here at Albenga in 1942. Powered by P.XI RC40 radial engines, the P.23R had taken the world speed record for 1,000 and 2,000 km distances in 1938. It was this aircraft which Nicolo Lana proposed to use to bomb New York.

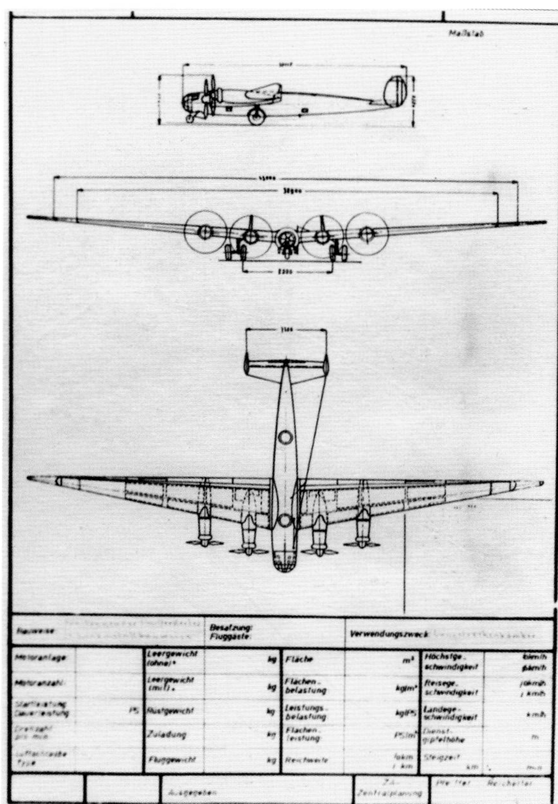

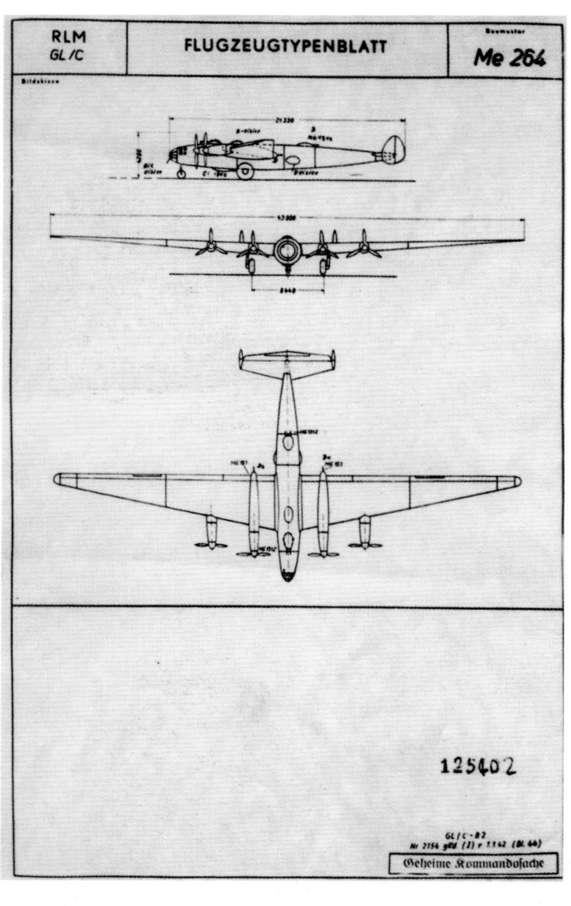
RLM GL/C

FLUGZEUGTYPENBLATT

Me 264

125402

Geheime Kommandosache

Left and bottom left The first known 'Flugzeugtypenblatt' – 'Aircraft Type Sheets' – for the Me 264 as issued by the RLM on 1 January 1942. In the Typenblatt to the left, the aircraft is seen in its basic state, while the sheet below shows the addition of in line engines and proposed armament. The aircraft is configured with remotely-controlled rearward firing 20 mm MG 151 machine guns mounted on the trailing edge of the wings and two upper fuselage turrets equipped with 13 mm MG 131 Zwilling units.

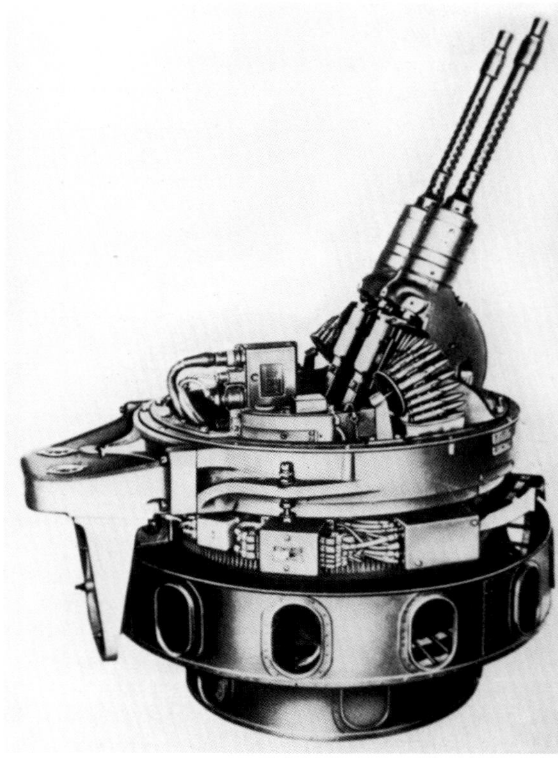

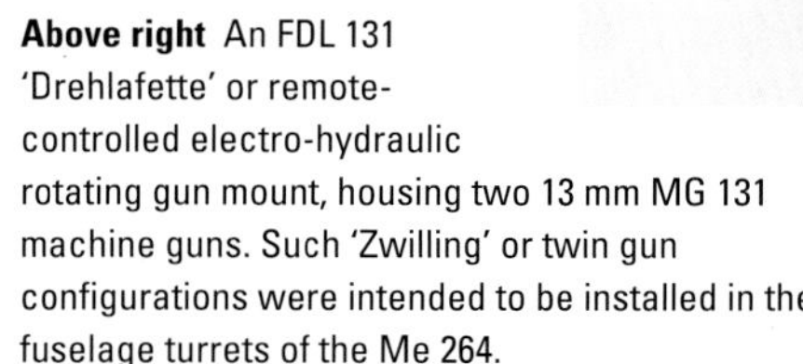

Above right An FDL 131 'Drehlafette' or remote-controlled electro-hydraulic rotating gun mount, housing two 13 mm MG 131 machine guns. Such 'Zwilling' or twin gun configurations were intended to be installed in the fuselage turrets of the Me 264.

Right An armourer crouches on the fuselage to remove a protective panel to the dorsal remotely-controlled MG 131 Z gun barbette as fitted to a Heinkel He 177. Behind the armourer can be seen the observation dome from where the gunner would have operated the guns. A similar weapons system was planned for the Me 264.

Below A page from an He 177 handbook showing the installation plan for a FDL 131/A MG 131 barbette – though this diagram shows a single-barrel mount as opposed to the 'Zwilling' configuration intended for the Me 264.

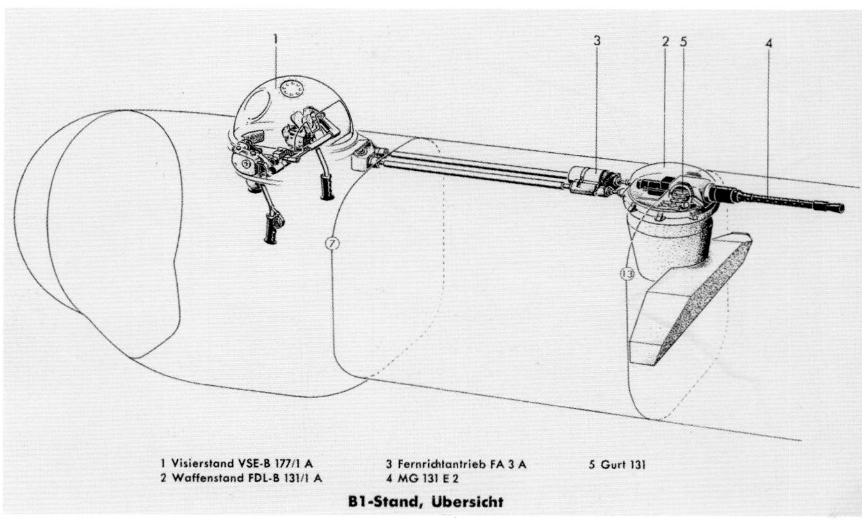

Above A jettisonable Walter 109-500 rocket unit of the type that would be fitted beneath the wings of the Me 264 to assist with take-off. Weighing at least 42 tons with a full fuel load, the Messerschmitt would require four such units which would provide extra thrust.

Lana proposed that by carefully conserving fuel, he would cut one engine as the fuel load gradually diminished, then another, then the third, making a bomb-run on just one engine. He would then make a water landing in the bomber in the Atlantic 200 km off Nantucket and clamber aboard a waiting Italian submarine. Although Lana – rightfully – saw his plan more as a propaganda mission rather than as having any serious strategic impact, the Air Ministry rejected it on the grounds that Lana's civilian status was at odds with the military nature of the mission. Matters were concluded however, when the Piaggio suffered a landing accident in a strong crosswind when piloted by Captain Max Peroli at Albenga on 23 May 1942.[28]

At Augsburg, the approach was, at least, more scientific. On 9 April, *Flugbaumeister* Scheibe of the RLM chaired a meeting to discuss the possibility of fitting the Me 264 with Jumo 211 engines as used in the Ju 88. The meeting was attended by a representative of the *Erprobungsstelle* Rechlin as well as by Woldemar Voigt, Paul Konrad, Karl Seifert, Hans Hornung and *Herr* Hugo of the Messerschmitt *Projektbüro*. With a projected all-up weight of 42.9 tons it was calculated that the Me 264 would have a range of 12,300 km at an altitude of between 2-3 km. The aircraft would need to be fitted with four auxiliary RATO units providing thrust of 1,000 kg, burning for 45 seconds each, over a take-off distance of 1,100 m. Armour protection planned at 2 tons would reduce range by 1,600 km to 10,700 km. It was planned to build the first two Jumo 211-powered prototypes – the V 1 and V 2 – without weapons and equipment. The V 3 was to be fitted with a bomb bay, weapons and also Jumo 211 engines.[29] A fourth prototype – the V 4 – was planned, and was to be the first 'V' machine to be fully equipped for an initial series production of 30 aircraft.

The meeting with Scheibe was the first of several dealing with the Me 264 to take place during April 1942 involving senior *Luftwaffe* officers.

On 14 April, *Generalstabs-Ingenieur Dipl.-Ing.* Roluf Lucht, Milch's senior engineer, complained to his boss that *Professor* Messerschmitt hampered his company's production by continually modifying his designs. An irritated Milch responded that he was seriously considering dismissing Messerschmitt from his position as head of his company so that he could be left to concentrate solely on design work. Lucht was promptly sent to Augsburg and Friedrich Seiler, Messerschmitt's deputy, was duly informed by Milch that Messerschmitt should step down and concentrate only on research.[30]

Clearly troubled by the Messerschmitt problem, the following day Milch called in the Chief of the RLM *Planungsamt* (Planning Office), his old colleague from his time at *Deutsche Lufthansa*, *Generalmajor* Carl-August *Freiherr* von Gablenz. Like Milch, von Gablenz had flown as a combat pilot during the First World War. An extremely accomplished aviator, he was also no stranger to the challenge of long-distance flight, having flown in a Ju 52 from Berlin-Tempelhof, via Belgrade, Athens, Cairo, Baghdad, Calcutta, Bangkok and Canton to Shanghai – a distance of 14,000 km – in August 1934. In 1936 he had made the 3,850 km flight from the Azores to New York. He had worked for Junkers as a technical assistant before joining *Deutsche Lufthansa* in 1924, rising to its Executive Board in 1933, a position he maintained until his death in an air crash in August 1942. Simultaneous to his role with *Deutsche Lufthansa*, he fulfilled positions within the *Luftwaffe*, commanding a *Blindflugschule* before being appointed *Lufttransportführer*. Milch instructed von Gablenz to despatch a *Luftwaffe* Commission to Augsburg to inspect and assess all working methods and plans associated with the Me 264. Milch was clear in his requirements:

> *'You are instructed to co-ordinate the Me 264 programme. This is to be done in conjunction with GL/C. You are authorised to form a commission from any person you consider suitable.*
>
> *'In particular, the following is to be determined:*
>
> Question 1: *Is there a chance that the type will fulfil the requirements as offered by Messerschmitt?*
>
> Question 2: *If so, at what dates for the individual aircraft?*
>
> Question 3: *What additional efforts, at present unknown to the RLM, are to be expected?*
>
> Question 4: *At what time can properly supported military operations with the promised effect begin?*
>
> Question 5: *Can the desired mission be solved more efficiently and more quickly by other means?'*

Below In April 1942, a study group consisting of representatives of the RLM, the Erprobungsstelle Rechlin and Messerschmitt at Augsburg, proposed the use of the Jumo 211 engine with at least the first three of the forthcoming Messerschmitt Me 264 prototypes. This 1,200 hp engine was fitted to the Ju 88 medium bomber and had proved itself a reliable powerplant.

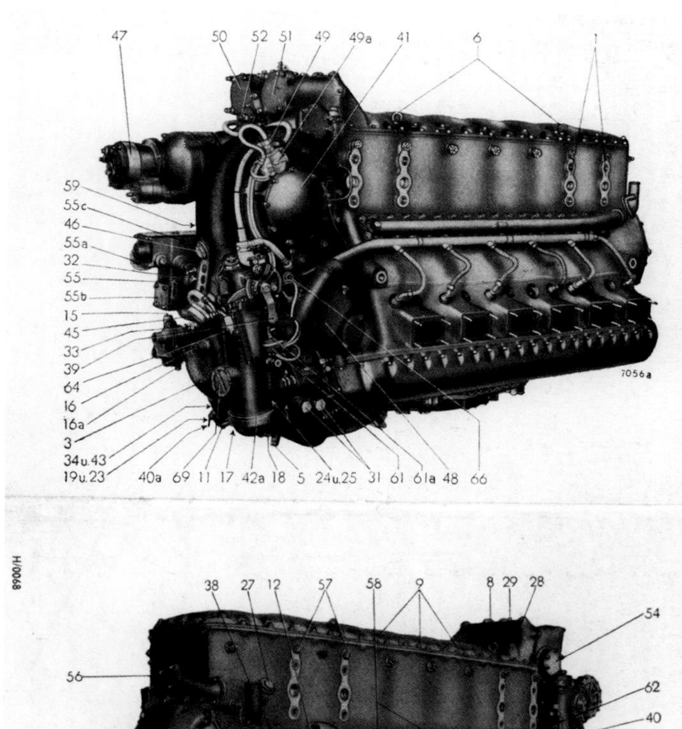

Above The locating and fixing point for the port wing on the fuselage of the Me 264 V 1, Augsburg, 1942 (see also page 55). Note also the crew access hatch just below the forward edge of the fixing point.

Below Close-up of the locating and fixing point for the port wing on the fuselage, showing connection braces.

Above Cross-section of the port side wing or tail assembly, prior to assembly to the fuselage. Note that the wing is resting upside down on a cradle.

Below Close-up of an assembly clamp fitted to the port side wing or tail assembly.

Below A section of wing under construction at Augsburg, 1942.

Below right Close-up of the brace applied to the wing section seen in the photograph below.

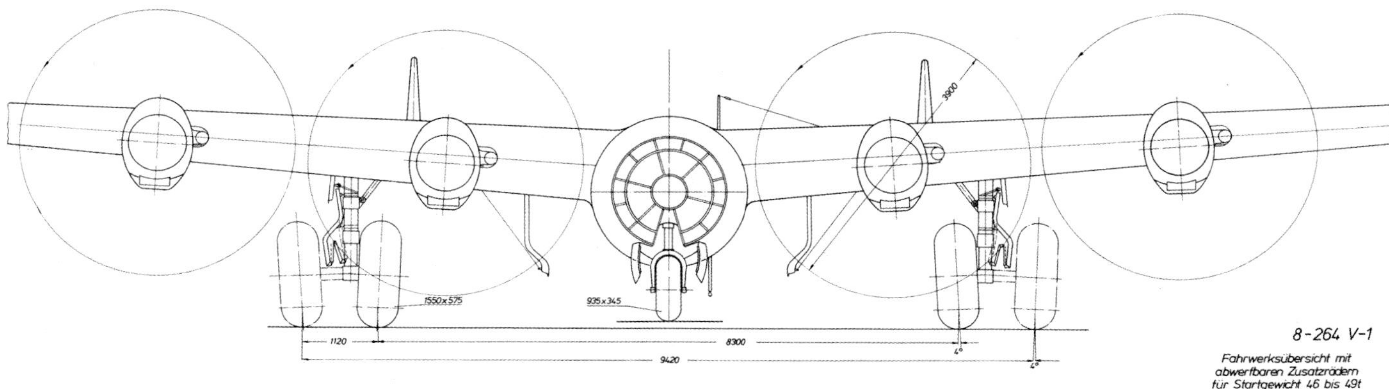

Above Based on a Messerschmitt drawing, this front view shows the aircraft fitted with Daimler-Benz engines which indicates that the diameter of the propellers on this version is 3,900 mm, whereas the propeller diameter on the Jumo-powered version would have been 3,630 mm. The drawing also shows the jettisonable outer mainwheels intended to support the take-off weight of the aircraft.

Prior to leaving for Augsburg, von Gablenz carried out some preliminary research; and responded to each of Milch's questions as follows:

'*The following points will answer the various questions*:

'Question 1: *With the Jumo 211 engine, the intended objective – nuisance raids against America – cannot be reached. Occasional missions with mid-air refuelling would be possible; however the aircraft would not be suitable for continuous* [non-stop] *operations.*

'Question 2: *Of the three V-aircraft* [prototypes] *under construction, the first could be finished after full resumption of work within 5 to 6 months. The two others would follow about two months later.*

'Question 3: *The departments concerned are aware of the labour effort required. However it is thought advisable to provide for a new cockpit – which has been demonstrated as a mock-up – from the 4th aircraft onwards, even if a delay would thus be caused. In this version, the crew seating arrangement has been considerably improved as compared to the original version. It is advisable however, not to delay the manufacturing process. In such a case, the change could be made at a later date (6th or 7th aircraft).*

'Question 4: *Properly supported military deployment with the expected results cannot be obtained with the Me 264. It is necessary, however, to have a number of aircraft in operation for other long-range missions, to the effect that the type can be employed operationally on occasion over the Atlantic with the object of armed reconnaissance, keeping contact with bomber units, cooperation with U-boats, the dropping of radio-location buoys behind convoys to facilitate location by bombers or U-boats and similar missions.*

'In view of the present non-availability of fast, long-range bombers, it is necessary to quickly allocate aircraft for such tasks. At the same time, as shown under question 5, the Me 264 is a step in the direction of future development.

'Question 5: *As pointed out in the foregoing replies, only a limited solution of the* 'Aufgabe Amerika' *is offered by the Me 264 in its present stage of development. So far, the other projects submitted do not offer an increase in range as compared with the Me 264.*

Below Unfortunately the date cannot be discerned on this revised RLM Typenblatt, which shows the proposed three wingspan variations as well as the additional, jettisonable outer mainwheels intended to support the take-off weight of the aircraft.

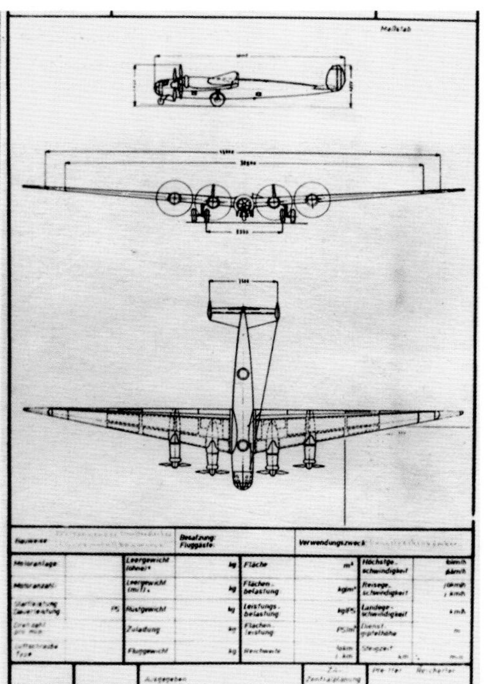

'The task of undertaking nuisance raids against the American East Coast demands so much from the design of the aircraft that it appears impossible at the moment to find a solution without an intermediate development stage. Therefore, the Me 264 represents a vital intermediary stage in development.

'From the designs so far submitted by various companies, only a new suggestion by Messerschmitt to increase the Me 264 to six engines promises to fulfil expectations. The other companies have designs in preparation which may represent the intermediate stage for the solution of the 'Aufgabe Amerika'.

'Junkers will shortly fly the Ju 290. The Ju 390, based on this development, does not meet the conditions required for the 'Aufgabe Amerika'.

'Currently, Heinkel has put forward the He 177, which at present is the aircraft with the greatest range. But a newly submitted design does not have the required range.

'Focke-Wulf, having obtained experience in developing the Fw 200 into the Fw 300 design, have also not been able to meet the necessary requirements.

'Under these circumstances, it appears advisable to continue development work on the Me 264, which has already progressed considerably, under all circumstances.'[31]

Von Gablenz' mention of Messerschmitt's proposed six-engined variant of the Me 264 was a reference to the P 1075, the ultra-long-range, six-engined version intended for operations over Asia, Africa and North America. With the total weight of fuel, armour, offensive load and armament still limiting range, so Messerschmitt accepted that the only swift and effective way to enhance the aircraft would be to enlarge the airframe and introduce two further engines, whilst keeping basic design and equipment elements unchanged. This would be accomplished by extending the fuselage by two additional frames and increasing the wing

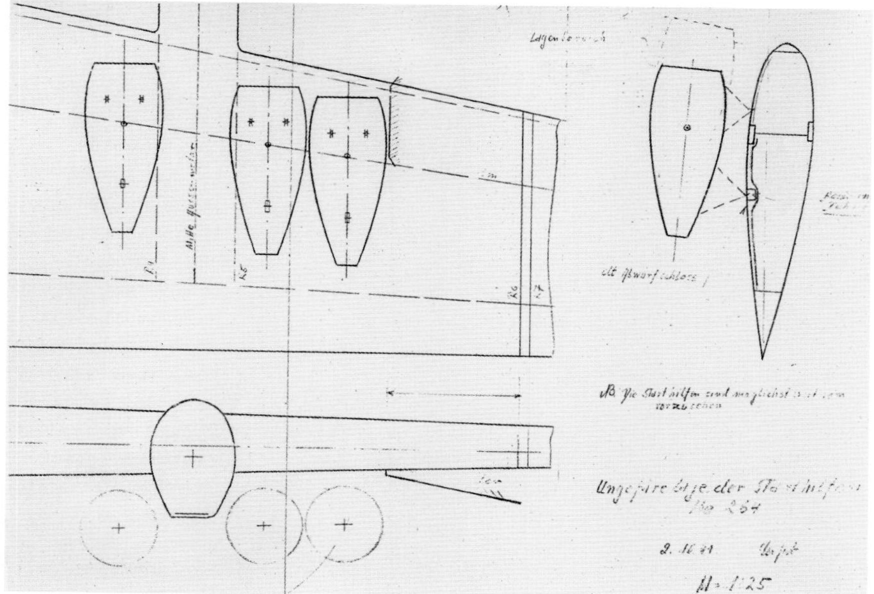

Left A drawing which appears to date from October 1941 in which three Walter RATO units are attached to the underside of the Me 264's wing on either side of the outer engine – in this case an unspecified Daimler-Benz type.

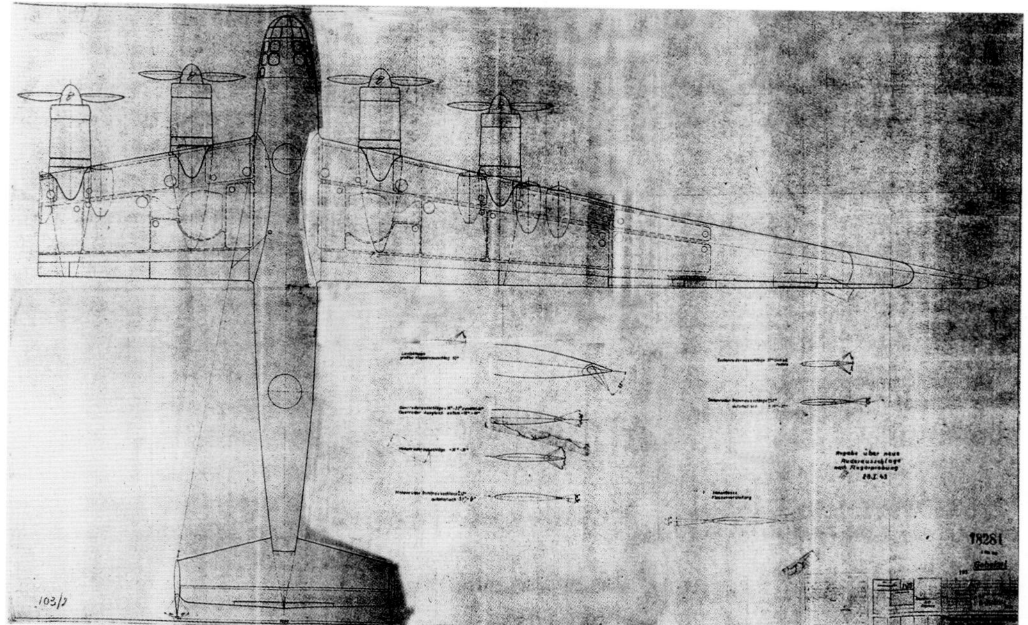

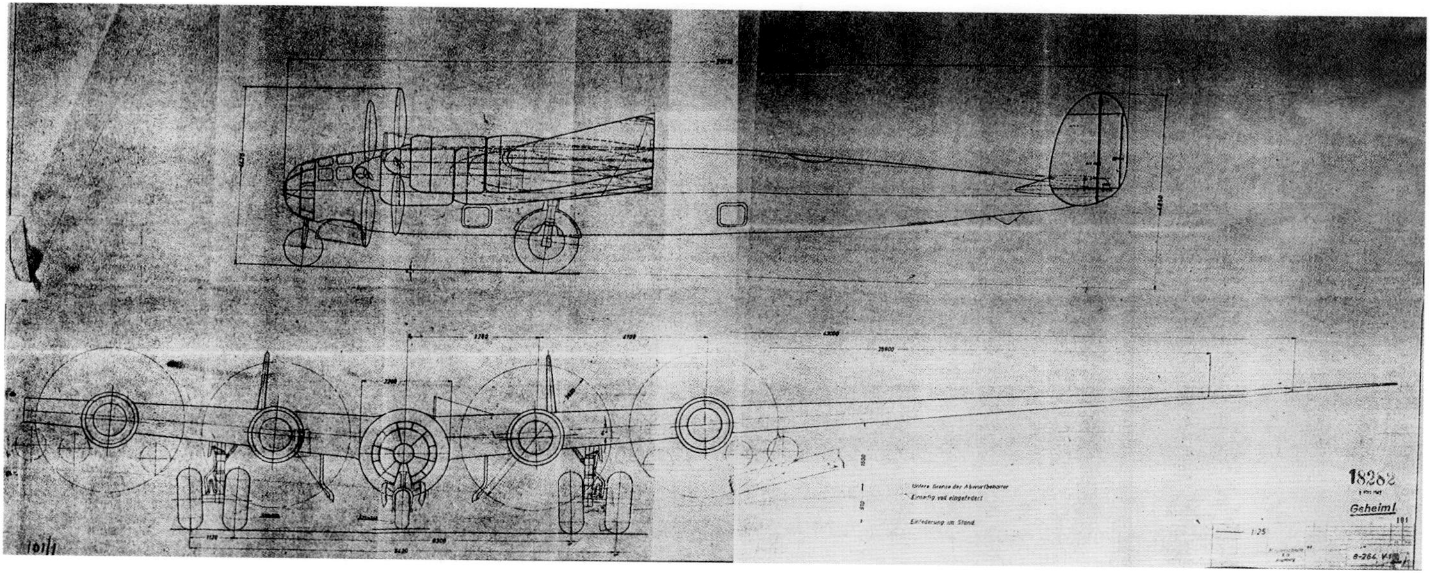

Left and below
Two Messerschmitt factory drawings from 28 January 1943 showing in more detail the proposed fitting of three Walter RATO units under each wing, together with the three planned wingspan variations and provisional positions for the fuselage uppersurface gun turrets.

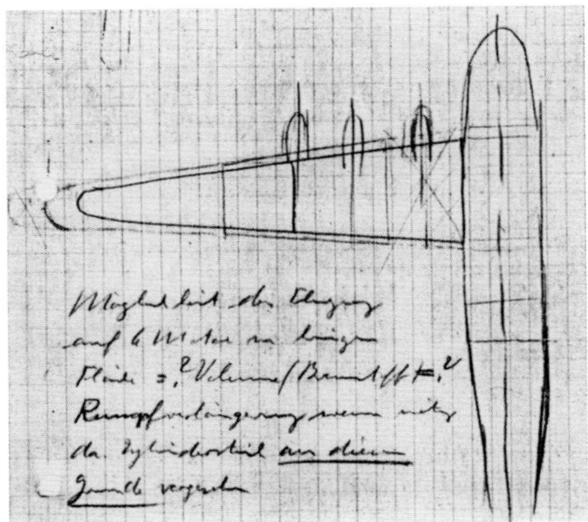

Above Professor Messerschmitt's sketch of a proposed six-engined Me 264, dated 16 February 1942. Messerschmitt has written: 'Possibility for an aircraft with six engines'. This was to be achieved by increasing the wingspan and fuselage length.

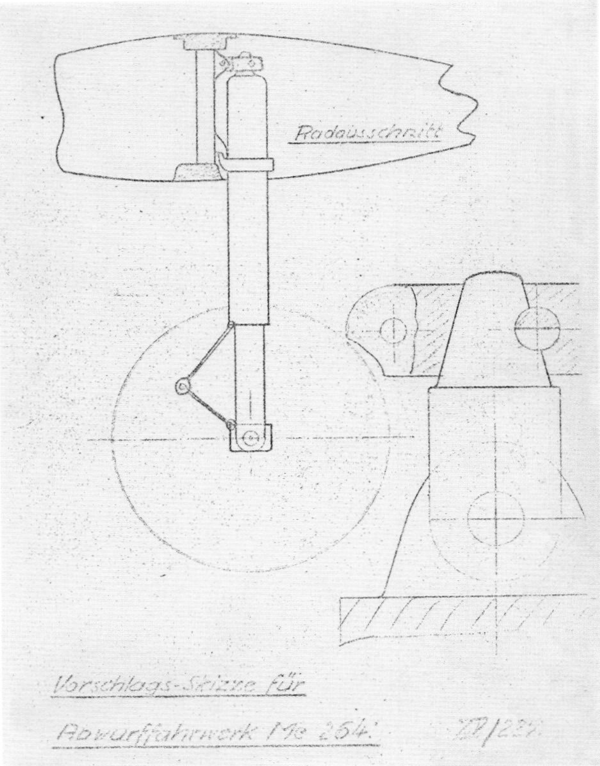

Left A provisional sketch produced by the Messerschmitt design office showing the proposed installation of the jettisonable outer single mainwheel and how it would be housed within the wing. It was intended that the wheel would be dropped after take-off and returned to the ground via parachute, although the storage for the latter is not shown.

Above and above right Two photographs showing the wind tunnel model of the proposed six-engined Me 264 with Jumo engines.

Right The four-gun HL 131 V rear turret equipped with four 20 mm MG 131 machine guns as fitted to a Heinkel He 177 and also proposed for the Me 264 (see drawings opposite).

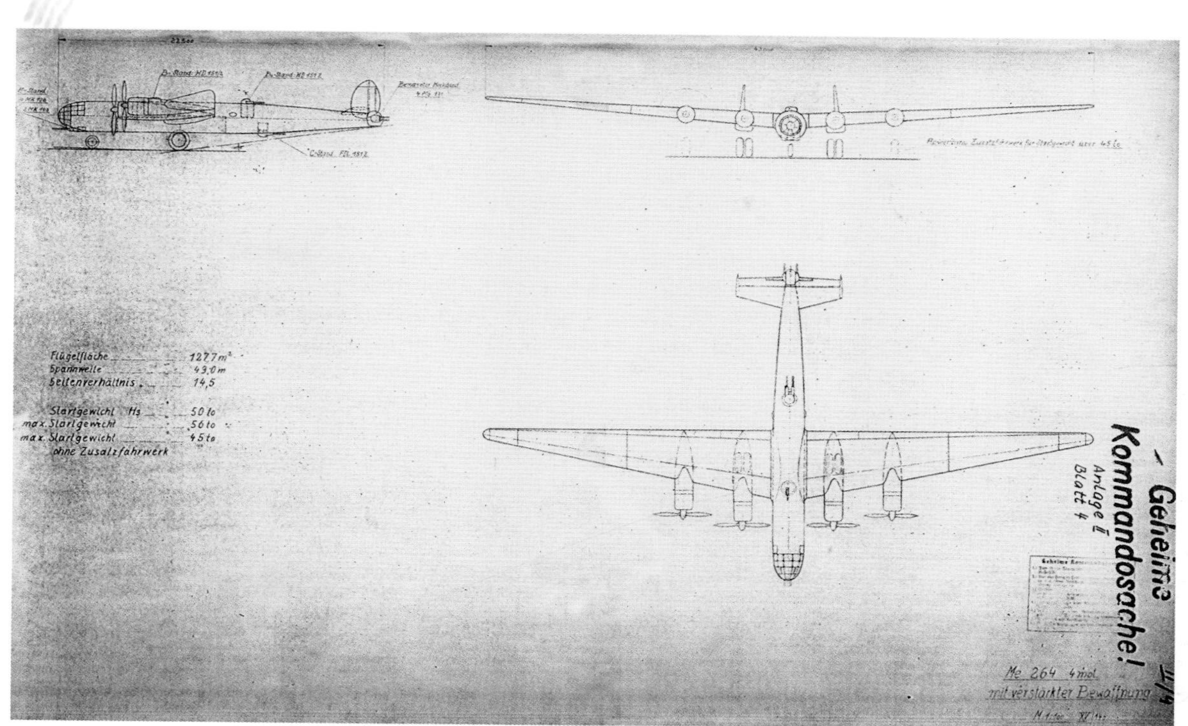

Above A three-view drawing dated 12 May 1943 showing the latest four-engined Me 264 configuration with proposed armament of four 30 mm MK 108 cannon or two MK 103 cannon in a nose gondola or chin turret. There are three turrets fitted with either HD or FDL 151 Zwilling (20 mm MG 151) gun mounts. A further rear turret was also planned with four MG 131 machine guns (see photograph on opposite page). Furthermore, the number of fixed mainwheels has been increased to four with two further jettisonable wheels under the centre engines. All-up weight was 45 tons, with a wingspan of 43 metres.

Below A three-view drawing dated 12 May 1943 of a six-engined Me 264 converted from a four-engined model, resulting in a 47.5 metre wingspan with a wing area of 170 square metres. This has been accomplished by increasing two sections of the fuselage length where marked and redesigning the inner wing sections to accommodate the extra engines. Additional flaps have been added to the trailing edge. Armament is as per the four engine drawing. All-up weight was 54 tons. It should be noted that the wingspan of the six-engined version is actually 4.5 metres longer than the four-engined version, as a result of the wing redesign. However, the overall length has increased from 22.5 metres to 25.5 metres.

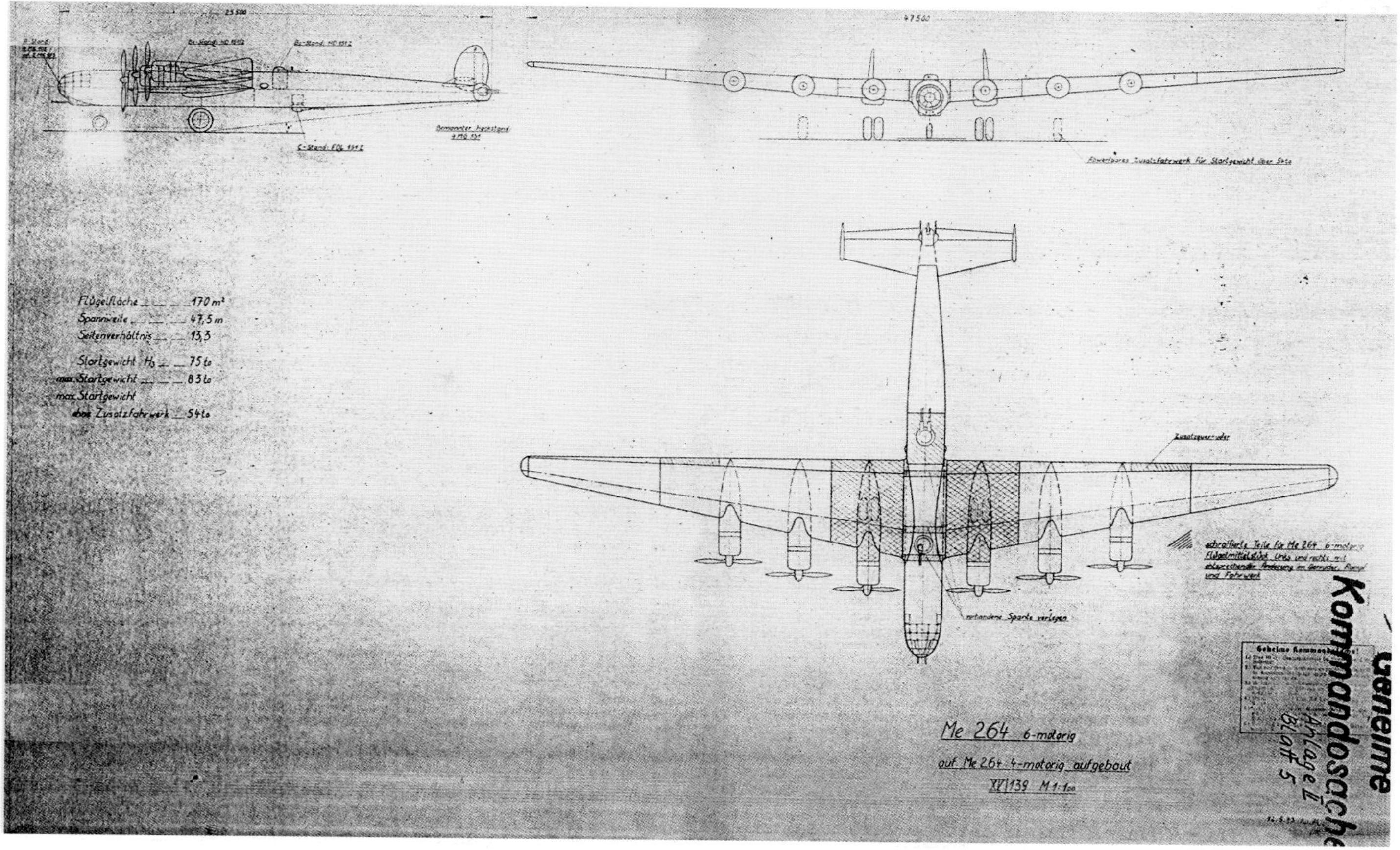

Generalmajor Carl-August *Freiherr* von Gablenz

In April 1942, *Generalfeldmarschall* Milch appointed his old colleague, *Generalmajor* Carl-August *Freiherr* von Gablenz, Chief of the RLM Planning Office, to act as inspector for the Me 264 project. He viewed the aircraft as '... a step in the direction of future development.' Born on 13 October 1893 in Erfurt, von Gablenz served as a pilot with *Feldflieger-Abteilung* 42, the *Kampfeinsitzerkommando* I and *Kampfgeschwader der OHL* (*Kagohl*) I during the First World War, with brief periods of service with the Staff of the Commanding General of the Imperial Air Service and as a Technical Office with the German flying contingent in Palestine between 1917-18. He joined Junkers in 1924 and then the airline, *Deutsche Lufthansa* (DLH), in 1926, with whom he worked until 1942, by which time he was on that company's executive board. Simultaneously to working with DLH, he also joined the Luftwaffe Reserve and served as both a pilot and observer. He joined the *Stab* of KGzbV 172 in September 1938 and was appointed commander of the *Luftwaffe Blindflugschule* (Blind Flying Schools) in November 1939. On 12 March 1940, von Gablenz became the '*Lufttransport Chef Land*' in charge of all land-based air transport units prior to operations over Scandinavia. From 5 May to 30 September 1941, he served as *Lufttransportführer*, responsible for all transport units, before being transferred to the RLM as *Chef des Planungsamts des Generalluftzeugmeisters* on 1 October 1941. He was killed in an air crash over Mühlberg on 21 August 1942. He was awarded the *Ritterkreuz des Kriegsverdienstkreuzes mit Schwertern* posthumously on 25 August 1942.

Above On 19 April 1942 Generalstabs-Ingenieur Dipl.-Ing. Roluf Lucht (left), a senior engineer with the RLM, arrived at Augsburg to assess progress on the Me 264. He had been sent there by Milch to investigate. Lucht subsequently reported that he found Messerschmitt to be a 'broken man'.

area to around 170 square metres. Whether the new engine power was to be provided by DB 603, DB 614 or Jumo 213 units was left open. Messerschmitt also proposed that in such a configuration, the aircraft be fitted with additional armament in the form of remotely-controlled MG 131 or MG 151 twin-gun turrets in the forward and rear fuselages.

The *Projektbüro* had completed work on the basic design by May 1942 which was supported by further options to allow a second machine to tow the main aircraft into the sky, mid-air refuelling and the fitting of up to six RATO units to conserve fuel during take-off. This latter configuration would have given the six-engined Me 264 a range of 18,000 km with a 5-ton bomb load or a maximum of 26,400 km without. [32]

On 19 April 1942, *Generalstabs-Ingenieur* Lucht arrived at Augsburg to inspect the state of work on the Me 264. Lucht found the atmosphere at the plant tense and reported to Milch that '... *I found Messerschmitt a broken man. He was physically at a very low ebb and crazy with emotion. He was crying like a baby*.'[33] Milch showed little sympathy and eventually Messerschmitt resigned as Chairman and Managing Director of Messerschmitt A.G., to be replaced by Theo Croneiss, a First World War flying veteran, Chairman of the Messerschmitt Shareholder's Committee and a man with close links to the Nazi Party. *Professor* Messerschmitt's activities would be restricted to that of Technical Director and Head of Development and he would play no part in the day-to-day affairs of the company. But Milch warned Croneiss and Seiler that unless the situation within the company did not improve he would consider appointing a '*Kommissar*' to run things.

Five days later the first element of von Gablenz' Commission arrived, headed by the *Ritterkreuzträger*, *Major* Edgar Petersen, the former *Kommodore* of *Kampfgeschwader* 40 who, late the previous year, had been appointed *Kommandeur der Erprobungsstellen*. Petersen was one of the *Luftwaffe's* foremost anti-shipping commanders with operational experience gained over Narvik, the North Sea and Atlantic flying the Ju 88 and Fw 200 *Condor*. He was accompanied by *Flugbaumeister* Scheibe of the RLM, *Stabs-Ingenieur* Harry Böttcher from the *E-Stelle* Rechlin, *Flugkäpitan* von Engel who had flown long-range flights into the North and South Atlantic during the 1930s, and *Oberingenieur* Schulz.

The Commission got to work quickly. On 25 April a meeting was held with *Professor* Messerschmitt, Croneiss, Voigt, Seifert, *Dipl.-Ing.* Hederer, Head of the Messerschmitt Technical Office and *Dipl.-Ing.* Gerhard Caroli, Head of Flight-Testing. The Messerschmitt team presented Petersen with the most recent blueprints, plans and performance projections for the first three Me 264 prototypes; the designs were virtually complete – it was only the wait to obtain DB 603 engines which was delaying progress. The Commission put forward the option of using Jumo 211 engines as an alternative, but the Messerschmitt representatives responded that the objective of reaching America, using an aircraft weighing 43 tons in such a configuration, would not be possible without refuelling.

Discussion then continued on the basis that BMW 801 engines could be used and Messerschmitt was asked to produce performance statistics accordingly. There was also further talk about a six-engined variant using Jumo engines, and extending the fuselage to increase tankage, which would bring the American East Coast within range.[34]

Later that day, *Generalmajor* von Gablenz arrived from Berlin. *Professor* Messerschmitt quickly took him to one side and presented him with a paper entitled '*The Me 264 on Atlantic Operations*'. It was a clever move, for it gave a new purpose to the Me 264 while avoiding the main objective of reaching the American coast when range remained a problem. Messerschmitt proposed enthusiastically that 'fan-like' formations of Me 264s should hunt enemy convoys far out in the Atlantic and then drop FuG 302 C '*Schwan*' radio buoys to guide Dönitz' U-boats towards their targets. Simultaneous to this reconnaissance role – for which it would also carry three Rb 50/30 cameras – the Me 264 would additionally strike at enemy warships and escorts using the latest anti-shipping weapons and be able to defend itself with an increased armament of MG 131 and MG 151 guns in turrets and lateral window positions.[35] Messerschmitt foresaw an aircraft with a flight duration of 45 hours, cruising at 350 km/h, though when equipped with GM-1 nitrous-oxide power booster units, it would be able to remain at 8,000 m for approximately 25 minutes at more than 600 km/h. By reducing bomb load to 2,000 kg, range would have been 11,000 km.

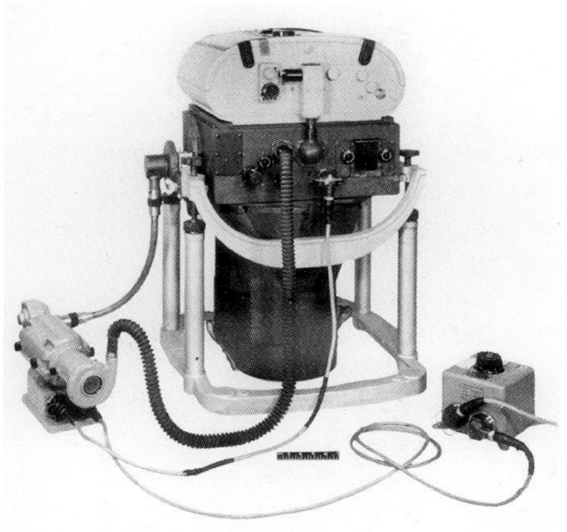

Left As part of his vision for the multi-role Me 264, in addition to operating as a bomber, Professor Willy Messerschmitt saw the aircraft fulfilling a long-range maritime reconnaissance role as support to the U-boats operating in the Atlantic. For such a role, the aircraft would be fitted with three Rb 50/30 aerial cameras, each weighing 725 kg, as seen here. The camera is mounted in a standard cradle; to the right is the film magazine which would normally contain 64 m of film, while to the left is the external motor.

Major Petersen's subsequent report dated 7 May estimated that the Me 264 could be available for operational missions against the United States by the autumn of 1943.[36] His conclusions were as follows:

'In its current configuration, the aircraft is designed as a four-engined long-range aircraft without pressurized cockpit, with bombs and armour as well as manually-controlled gun stations.

Engines are either 4 x 801D or 4 x Jumo 211J.

Using the 211J powerplant, estimated range is 13,000 km; with the 801D it is 14,000 km. Take-off weight is 45 metric tons in both cases. Thus, the east coast of America is just within range.

A 2,000 kg bomb load is possible if the armour and window guns are dispensed with.

With regard to the performance, using the 801D the aircraft just meets the specified requirements. The airframe's suitability for overloaded take-offs can only be determined after evaluation.

The first prototype will be ready to fly in the autumn of this year, with V 2 and V 3 following sometime during the winter. Full-scale production would get underway in the winter of 1942/43, with operations using single aircraft possibly beginning in the autumn of 1943.

We know of no other aircraft which has the same range with a similar speed and payload.

In conclusion, the consensus is that it would be worthwhile to build the project, submitted by Messerschmitt A.G. … and utilize the type for Atlantic operations.

Regardless of whether the type would be able to reach the American east coast in all circumstances there is still an urgent need for a long-range aircraft with the ability to scout the Atlantic as an armed reconnaissance platform and act as a reconnaissance aircraft and convoy shadower for He 177 formations.

Other important roles for such an aircraft include:

Joint operations with U-boats

Shadowing convoys

Dropping radio buoys behind convoys to plot their locations for bombers and U-boats

Long-range maritime reconnaissance operations

In view of the current shortage of high-speed bombers it is recommended that a limited number of aircraft be built with the caveat that the aircraft should be employed operationally only if it is fitted with the new all-round canopy glazing and instrument suite.'[37]

Petersen's report effectively gave Messerschmitt the green light to continue work, and Milch also begrudgingly authorised work to continue on the Me 264 at the beginning of the summer – if nothing else, the *Generalluftzeugmeister* recognised the propaganda value of long-range harassment missions against the United States.[38]

By mid-1942 the war had become global. Ready for the fight in the Atlantic, the *Kriegsmarine* was preparing to sail the first of its submarine tankers, laden with stores, spares and diesel fuel which would supply its packs of U-boats operating off the US East Coast and in other distant waters such as the Gulf of Mexico and the Caribbean Sea. In Germany itself, the British bomber offensive was inflicting severe damage. On the night of 28 March the target had been the Baltic city of Lübeck, where 2,000 buildings were destroyed and 312 civilians killed, with another 15,000 left homeless.

At the *Erprobungsstelle* Rechlin, former test-pilot *Oberst-Ingenieur Dipl.-Ing.* Dietrich Schwencke, who was quietly working away from the limelight on a very detailed report for his superior, Erhard Milch, was called from his work to examine the wreckage of an RAF Wellington bomber shot down that night. Amongst its smouldering remains he made an important discovery, namely the navigational equipment the British knew as 'Gee' which aided their pinpoint bombing attacks. It was another 'coup' for Schwencke.

Dietrich Schwencke had served as an assistant air attaché to London in the 1930s before being employed at Rechlin since at least

Below Major Edgar Petersen (left), the Kommandeur der Erprobungsstellen and the former Kommodore of Kampfgeschwader 40, was a key member of the von Gablenz commission despatched to Augsburg by Generalfeldmarschall Milch to report on the technical and tactical possibilities offered by the Me 264. Petersen concluded that: '…the consensus is that it would be worthwhile to build the project, submitted by Messerschmitt A.G. … and utilize the type for Atlantic operations.' Petersen is seen here in discussion with Generalmajor Adolf Galland, the General der Jagdflieger, at Lechfeld in 1943.

In April 1942, Oberst-Ingenieur Dietrich Schwencke, an experienced RLM engineer attached to the Erprobungsstelle Rechlin, produced a detailed report for Göring on the various types of long-range aircraft available to the Luftwaffe and their operational capabilities. Within the same report, Schwencke also listed 21 potential targets in North America for long-range bombers including aluminium factories, aero-engine plants and instrument and optical glass factories in the New York and New Jersey areas as well as others across New England and the Mid-West.

1934 when he had headed up *Gruppe* T3 which oversaw matters related to mechanical, W/T, photographic and navigational equipment.[39] He then became involved in the salvaging and assessment of captured foreign aircraft and aeronautical equipment. During the German invasion of France in 1940, Schwencke collected large numbers of captured French and British aircraft and sent them to Rechlin for examination. He worked closely with the military intelligence services and in March 1941, three months before the German invasion of the Soviet Union, he was sent to Russia to assess the strength of the Soviet Air Force as well as Russian aircraft production capacity. Later, in the summer of 1942, he was to conduct one of the first examinations of a shot down American B-17 Flying Fortress and made detailed reports for Göring and Milch on his findings. He even secured a work force of 200 Russian PoWs whose task it was to cut up salvaged enemy aircraft for technical evaluation.[40]

In April 1942 Schwencke had been asked by Milch to produce a thorough assessment of the various types of long-range aircraft available to the *Luftwaffe* and to outline the various missions to which they could be assigned. His study was to be made for the personal attention of no less than *Reichsmarschall* Göring. He completed his report on 27 April, the day that found Hitler in a furious mood following a raid by RAF Bomber Command against Rostock which had destroyed 70 per cent of the heart of the old city and left the Heinkel works there badly damaged. The *Führer* had told Josef Goebbels, his Propaganda Minister, that he would continue the recent German raids against British cities '... night after night until the English were sick and tired of terror attacks.' That night RAF Bomber Command bombed Rostock for the second time in three days, causing widespread destruction and forcing 100,000 citizens to be evacuated.

The 33 page report which eventually found its way to Göring's desk, entitled '*Einsatzaufgaben für Fernstflugzeuge*' ('Operational Tasks for Long-Range Aircraft')[i], was extremely comprehensive and detailed.[41] Schwencke broke his report down into key areas: firstly, the movement of vitally needed commodities such as tin, rubber, copper, molybdenum, tungsten and platinum as well as military equipment from Japan and Japanese controlled areas of Asia to Germany – though Schwencke noted that certain items could be obtained from South American countries such as Argentina, Chile, Bolivia, Brazil and Colombia; secondly, the potential for long-distance flights between 1942-44 and their likely capacity as well as an outline of reconnaissance objectives and military targets in the Soviet Union; thirdly, similar flights to central and north-east Africa and the Persian Gulf; and fourthly, to the east coast of North America (what Schwencke called the 'Western Atlantic area').

Schwencke concluded that there were five types of aircraft capable of undertaking one, or several, or all of the aforementioned requirements: the Heinkel He 177 (using mid-air refuelling); the Focke-Wulf Fw 200 (unarmed), the Blohm & Voss BV 222 (a heavier, 50-ton variant using mid-air refuelling); the Junkers Ju 290 (using mid-air refuelling) and the Messerschmitt Me 264 (with options for Jumo 211 or DB 613 A engines).

The supply route between Germany and Japan was seen as Berlin-Petsamo in Finland (2,100 km) / Petsamo-Tsitsihar in Central Manchuria (5,650 km) / Tsitsihar-Nagasaki (1,650 km), while a further route was planned for Berlin to Penang via Constanza on the Black Sea (1,600 km) and then Constanza to Penang (7,800 km) overflying Iran, Afghanistan and India.

Potential military targets in the Soviet Union were seen as petroleum refineries, synthetic rubber plants, aluminium and iron works, smelting plants, and aircraft, armaments, munitions, vehicle and military machinery plants including those at Baku, Chalilovo, Irkutsk, Komsomolsk, Kazan, Novosibirsk, Orsk, Stalinsk and Ufa.

Schwencke also saw German long-range aircraft as undertaking reconnaissance and bombing missions over Dakar, Conakry, Freetown, Abidjan, Lome and Lagos as well as Cairo, Port Sudan, Aden, Port Said and Suez – all points in Africa where the enemy, particularly the British, had a military, maritime and/or naval presence or where they had vital supply interests. The report also quoted the opportunities to be had in aircraft being able to reach Iraq (Basrah), Iran (Bander-Shahpur) and the Persian Gulf (Bahrain).

In dealing with the United States, Schwencke noted that American aluminium works and aero-engine plants, propeller factories and armament plants could be attacked only by the Me 264 fitted with DB 613 engines, carrying a 5,500 kg offensive load and operating out of Brest in western France. On the other hand, if the Azores could be used as a transit landing ground and refuelling point, then it would be possible to reach US targets with the He 177 (5 ton bomb load and refuelled), the BV 222 (4.5 ton bomb load and refuelled), Ju 290 (5 ton bomb load), and the Me 264 (increased 6.5 ton bomb load).

Schwenke's report listed no fewer than 21 potential targets in North America and Greenland including aluminium and aero-engine plants, equipment and armament plants. They were:

i. In his book on Germany's military efforts towards the United States during the Second World War, *Target: America – Hitler's Plan to Attack the United States* (Praeger, 2004), author James P. Duffy credits historian Olaf Groehler with the discovery at the Military Archives in Potsdam of the only known copy of the Schwencke report, prior to which '...many historians and researchers doubted the existence of such a document.' For the record, a copy of this document was located a number of years prior to the publishing of the book you now read, on microfilm, within the US Air Documents Division T-2 collection at the Imperial War Museum, London, by aviation historian, J.Richard Smith. The document, '*Einsatzaufgaben für Fernstflugzeuge*', [GL/A-Rü/Br. 208/42, 12 Mai 1942], a copy of which is in this author's possession, can be found on microfilm Reel 3623, commencing at Frame 16. The author is grateful to J.Richard Smith for making him aware of the existence of this document in the IWM archives.

The Schwencke document is also quoted by Karl Kössler and Gunther Ott in their excellent history of the Junkers Ju 89-Ju 390 series, *Die großen Dessauer – Junkers Ju 89, Ju 90, Ju 290, Ju 390 – Die Geschichte einer Flugzeugfamilie*, published by Aviatic Verlag, Planegg, in 1993, pg 96 and note 135.

Company	Product	Location	State
Aluminium Comp. of America	Aluminium	Alcoa	Tennessee
Aluminium Comp. of America	Aluminium	Massena	New York
Aluminium Comp. of America	Aluminium	Vancouver	Canada
Aluminium Comp. of America	Aluminium	Badin	North Carolina
Wright Aeronautical Corp.	Aero-Engines	Paterson	New Jersey
Pratt & Whitney Aircraft	Aero-Engines	East Hartford	Connecticut
Allison Div. of Gen. Motors	Aero-Engines	Indianapolis	Indiana
Wright Aeronautical Corp.	Aero-Engines	Cincinnati	Ohio
Hamilton Standard Propellers	Aircraft Propellers	East Hartford	Connecticut
Hamilton Standard Propellers	Aircraft Propellers	Pawcatuck	Connecticut
Curtiss Wright Corp Prop. Div.	Aircraft Propellers	Beaver	Pennsylvania
Curtiss Wright Corp Prop. Div.	Aircraft Propellers	Caldwell	New Jersey
Sparrow Gyroscope Co.	Instruments & Searchlights	Brooklyn	New York
Cryolite Mine	Cryolite	Arsuk	Greenland
Cryolite Refinery	Cryolite	Natrona (Pittsburg)	Pennsylvania
American Car & Foundry	13 ton tanks	Berwick	Pennsylvania
Colts Patent Fire Arms Mfg. Co.	37 mm AA guns & MG	Hartford	Connecticut
Chrysler Tank Arsenal	28 ton tanks	Detroit	Michigan
Allis Chalmers	90 mm AA guns	La Porte	Indiana
Corning Glass Works	Optical glass	Corning	New York
Bausch & Lomb	Optical glass	Rochester	New York

Below Oberst-Ingenieur Dietrich Schwencke's target listings from his April 1942 report to Göring, together with a global map showing planned and existing reconnaissance, bombing, transport and courier flight paths. The numbers on the map correspond to the potential targets. Note the various types of aircraft – BV 222, He 177, Me 264, Ju 290 – are also indicated on the flight paths.

Wichtige Rüstungsanlagen an der Westatlantikküste und in USA.

Nr.	Firma	Erzeugnis	Ort	Staat	% der Gesamterzeugung
201	Aluminium Comp. of America	Aluminium	Alcoa	Tennessee	30
202	Aluminium Comp. of America	Aluminium	Massena	New York	15
203	Aluminium Comp. of America	Aluminium	Vancouver	(Kanada)	20
204	Aluminium Comp. of America	Aluminium	Badin	North Carolina	12
205	Wright Aeronautical Corp.	Flugmotoren	Paterson	New Jersey	16x
206	Pratt & Whitney Aircraft	Flugmotoren	East Hartford	Connecticut	20x
207	Allison Division of Gen. Mot. Corp.	Flugmotoren	Indianapolis	Indiana	10
208	Wright Aeronautical Corp.	Flugmotoren	Cincinnati	Ohio	15
209	Hamilton Standard Propellers	Luftschrauben	East Hartford	Connecticut	15x
210	Hamilton Standard Propellers	Luftschrauben	Pawcatuck	Connecticut	15
211	Curtiss Wright Corp. Propeller Division	Luftschrauben	Beaver	Pennsylvania	15
212	Curtiss Wright Corp. Propeller Division	Luftschrauben	Caldwell	New Jersey	15
213	Sperry Gyroscope Co.	Geräte, Instrumente, Scheinwerfer	Brooklyn	New York	50x

x Zugleich Entwicklungswerke.

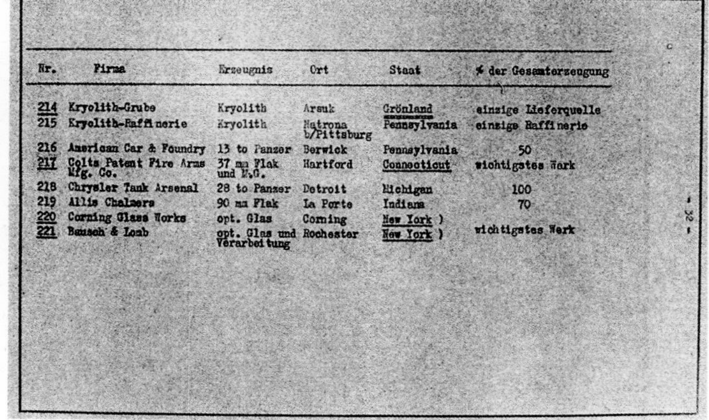

Nr.	Firma	Erzeugnis	Ort	Staat	% der Gesamterzeugung
214	Kryolith-Grube	Kryolith	Arsuk	Grönland	einzige Lieferquelle
215	Kryolith-Raffinerie	Kryolith	Natrona b/Pittsburg	Pennsylvania	einzige Raffinerie
216	American Car & Foundry	13 to Panzer	Berwick	Pennsylvania	50
217	Colts Patent Fire Arms Mfg. Co.	37 mm Flak und M.G.	Hartford	Connecticut	wichtigstes Werk
218	Chrysler Tank Arsenal	28 to Panzer	Detroit	Michigan	100
219	Allis Chalmers	90 mm Flak	La Porte	Indiana	70
220	Corning Glass Works	opt. Glas	Corning	New York)	
221	Bausch & Lomb	opt. Glas und Verarbeitung	Rochester	New York)	wichtigstes Werk

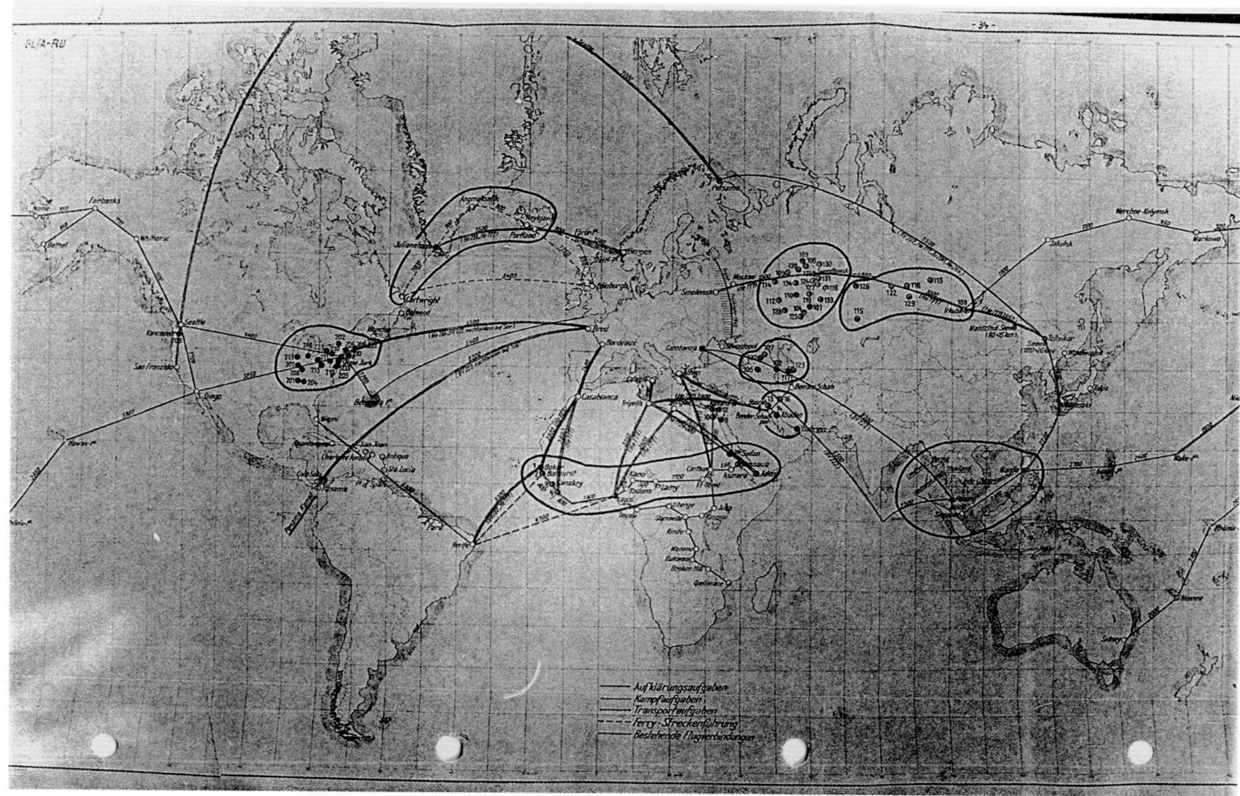

Schwencke concluded: '*A study of the armaments industry and the important raw material resources in the western hemisphere results in an indication of limited success against the 21 listed industrial targets. With this list, it should be asked how many of these targets will suffer (from bombing), how many will be affected by production delays, how many of the targets share production and what is the current supply situation...*

'*It is especially worth mentioning that the refining and output of natural cryolite comes exclusively from Greenland and is used in the production of aluminium.*

'*Concerning the condition of power stations, it is concluded that:*

In the USA, the construction of power stations has been greatly encouraged by the Government.

Each power station is linked to another by an extensive relay network.

The failure of 4 to 5 of the large power stations (from bombing) will have little effect on armaments production since the energy demand for civilian production exceeds the possibility for worthwhile reductions if required.'

In the United States itself, there was growing concern – indeed, acceptance – at the highest level that the country was no longer immune from the threat of bombing by an enemy power. On 18 April, in conditions of great secrecy and under the command of Colonel James H. Doolittle, 16 B-25 bombers took off from a US Navy carrier and flew more than 800 miles across the Pacific to attack targets on the Japanese mainland, including Tokyo, Kobe, Yokohama and Nagoya. Damage was inflicted to a number of oil and naval installations as well as an aircraft carrier at Yokusuka, although in strictly military terms the raid was a mere pinprick. However, that was not the point. The real result was that it left the Japanese shocked, angry and feeling vulnerable. The raid meant that the Japanese home islands were no longer safe and that was a blow to fighting morale. Conversely, in the United States, morale was greatly boosted, the American bomber crews were feted as heroes and the propaganda value was considerable.

However, behind the euphoria, one man at the heart of the American war effort was uneasy; Henry L. Stimson, the US Secretary of War, was quoted as viewing a reprisal strike against America by the Japanese '...inevitable'. Furthermore, Stimson warned that it would not be just the West Coast that was under threat; he told journalists on 28 May 1942 that in response to the attack on Tokyo '... an attack on Washington is not wholly inconceivable.' The *New York Times* reported that '... extraordinary precautions are being made to build up defences'.[42]

Above Following the success of the 'Doolittle Raid' against targets in Japan, in mid-1942 there was growing concern in the United States about the possibility of Axis bombing raids against American cities and precautionary measures were taken. Here, newly manufactured P-51 Mustangs at Mines Field, Los Angeles, home to the North American Aviation Company, await delivery to the USAAF beneath a barrage balloon intended as defence against low-flying enemy intruders.

In Germany, however, the possibility of mounting early operations against any US target was unlikely; following the information prepared by *Major* Petersen in early May 1942, those with a realistic view within the RLM worked on the basis that the Me 264 would not enter operational service until very late 1943 or most probably early 1944.[43] On 12 May 1942, a meeting took place in Berlin attended by *Generaloberst* Jeschonnek, *Generalfeldmarschall* Milch, *Generalmajor Freiherr* von Gablenz and various senior *Luftwaffe* officers involved with the Me 264 programme. The meeting included a number of presentations, one of which was by *Flugbaumeister* Friebel. Friebel played down Petersen's optimistic assessment of a few days earlier and spoke of '1944' as being a much more accurate time frame for commencement of long-range operations with the aircraft, by which time more sophisticated weaponry would be available in the form of long-range aerial torpedoes, remote-controlled bombs and rockets such as the PC 1400 X, and even smaller parasite defence fighters intended to be carried by the Me 264 'mother' aircraft. In the meantime however, Friebel did foresee that one or two prototypes would be ready for flight-testing in 1943.

In addition von Gablenz presented Jeschonnek with a 2,500 word document on the development of trans-oceanic aircraft. In von Gablenz' view '... *the design* [of the Me 264] *was so much better than its competitors as regards weight and performance that it was subjected to a lengthy evaluation before final acceptance. Because Messerschmitt lacks experience in building heavy aircraft, and also because in the opinion of the RLM technical specialists, the Me 264 plan-form was too narrow, the design could not be adopted as the standard long-range bomber for the* Luftwaffe *and other firms had to be invited to tender.*'

Von Gablenz outlined the various projects from Heinkel, Focke-Wulf, Blohm & Voss and Junkers and went into more detail on the Me 264:

'*Me 264 first variant:*	*All-up weight 50 tons, four DB 603 engines, range 13,000 km, payload 3 tons, 2 tons armament and armour.*
'*Me 264 second variant:*	*All-up weight 47 tons, four BMW 801 engines, range 12,000 km, payload 2 tons as reconnaissance aircraft, 3 tons armament and armour possible.*
'*Me 264 third variant:*	*All-up weight 43 tons, four Jumo 211 engines, range 11,500 km, no bombs, 1.4 tons armament and armour.*

'Investigation has shown that contrary to the assertion made by the firm [Messerschmitt], *the America return flight direct is probably not possible with so little reserve. However, compared with the other proposals, the lesser all-up weights and materials requirement, plus four engines as against five or six, the design is so much better that in the opinion of the* Generalluftzeugmeister *this is the aircraft to order and accordingly attempts will be made to resolve the difficulties of under-capacity at Messerschmitt.*

If it is possible to resume the interrupted course of development work on the Me 264, then we can have the prototypes this year, get the programme of flight testing over with during 1943, and have the machines operational in 1944. The simplifications to the design suggested by the RLM Technisches Amt *(normal unpressurised cockpit, proven engines, no bomb-aimer's position) were made so that development could be continued despite the under-capacity at Messerschmitt (which would otherwise involve very long delays in production or cancellation of the contract) and to get the aircraft into the air. Furthermore, on account of the increasing delay in trials and introduction of the DB 603 engines, it is not expected that the engines will be ready for the 30-hour endurance run in 1943/44.*

Even the Me 264 will need mid-air refuelling on the American route, but this is possible with the alternative engines. Particular value is placed on this assurance as, according to the Generalluftzeugmeister *Staff at a meeting with the Chief of the General Staff, it was assumed by the* Generalluftzeugmeister *that fitting alternative engine types and simplification of the Me 264 would mean giving up the idea of trans-oceanic operations.'*

'Me 264 six-engined variant: All-up weight 70-80 tons, six DB 603 engines, range 15,000 km, payload 5 tons, approximately 4 tons armament and armour.

'The solution was theoretical because of the firm's under-capacity. Mention must be made of the special characteristics associated with these trans-oceanic aircraft and their performance. They require long runways (on average a 2 km run before take-off). The aircraft are fitted with landing gear; a jettisonable undercarriage is provided for take-off. Tactically, problems of stability and lack of manoeuvrability rule out dive-bombing. For nuisance raids against American land targets, night bombing will be in the horizontal attitude. In operations over the sea, the size of the aircraft is particularly disadvantageous. Normal aerial torpedoes are envisaged, but remote-controlled missiles would be best. Planning and development work is in hand to launch small parasite aircraft from these very large aircraft to carry out attacks. Test aircraft of this type are under construction and the basic trials regarding mounting them on the mother aircraft have been partly completed and should be finished during the course of the year. The parasite aircraft will be very small (6 square metres as against the 14 square metres of the Bf 109) jet-propelled single-seaters designed for flights lasting from 30 minutes to an hour. They should be especially suitable for direct attacks on shipping using one or two bombs of 1,000 kg aggregate weight. Experiments to solve the problem of how the mother aircraft can recover the parasite are also in hand. Besides bombing, the parasite aircraft can be used to defend against enemy fighters. A Ju 390 for example can carry two parasite aircraft. The aircraft can be re-armed with bombs aboard the mother aircraft. To summarise therefore, in accordance with the suggestions of the Generalluftzeugmeister *in the area of trans-oceanic aircraft, the following have been approved:*

Below and bottom By mid-1942, planning and development work was underway to launch small parasite aircraft to carry out attacks on American cities from very large carrier aircraft, such as the six-engined Ju 390. In discussions with Milch on 12 May 1942, von Gablenz stated that for distances up to 10,000 km, the planned Ju 390 would be 'better' than the proposed six-engined Me 264 for carrying two such parasite fighters. The aircraft seen here is the Ju 390 V 1 (first prototype), with its six BMW 801 D engines and impressive 50 metre wingspan – eight metres greater than the Ju 290. The Ju 390, which was also six metres longer than the Ju 290, was constructed using many of that aircraft's components. The V 1 first flew from Junkers at Dessau in August 1943.

The four-engined Me 264 will be proceeded with as the quickest possible solution for operations against the United States.

The necessary mid-air refuelling procedure will probably be worked out during 1942.

For distances up to about 10,000 km, the Ju 390 is better (heavier loads, better armament, parasite aircraft.

Investigations with the object of achieving a return flight to and from America without refuelling (six-engined Me 264) will be stepped up.

Use of the Ju 290 for distances up to 8,000 km will be investigated with a view to using the aircraft to refuel the four-engined Me 264.' [44]

According to the memoirs of the *Luftwaffe* bomber pilot, Werner Baumbach, Milch '... *asked Jeschonnek the vital question: "Don't the General Staff really believe that by refuelling from the air at night we can send a bomber vast distances, probably several thousand kilometres from its base?"*

"There's no object in doing so!" was Jeschonnek's quick and peremptory reply.

'When I urged, as I frequently did, that such refuelling was the solution of all the difficulties surrounding the American project, Jeschonnek used to say: "I agree with you – but there's Russia!"'

'It was no use thinking of even the modest operation against the American continent while Russia was devouring all our available resources in aircraft.' [45]

The references by both von Gablenz and Friebel to 'parasite fighters' for attack and defence may have been what prompted Milch to begin discussing a plan with his close aides later that month. If an Me 264 could at least approach New York, or possibly even San Francisco, carrying a smaller aircraft, this smaller aircraft could then detach itself from the Messerschmitt 'mother', fly the remaining distance, drop – with some accuracy – a bomb on Manhattan, before ditching in the Atlantic where its crew would be picked up by a U-boat. [46]

The Argus pulse-jet powered Me 328, the cheap, quick-to-build and expendable air-launched fighter and fighter-bomber which it was proposed would be carried beneath long-range aircraft such as the Me 264 and Ju 390. Such 'parasite' aircraft were intended to carry bomb loads of up to 1,400 kg. In March 1943, the glider manufacturer, Jacob Schweyer Segelflugzeugbau, was commissioned to work with the DFS and Messerschmitt to build the Me 328. The firm produced a wooden, mid-wing aircraft, with a circular-section fuselage featuring a raised cockpit canopy faired back and down to the base of the fin. The wingspan could be adjusted by means of detachable tips and electrical power was provided by wing-mounted air-driven generators. Self-sealing fuel tanks were housed in the rear fuselage and nose. Landing was to be accomplished using a retractable skid and the pilot was protected by a bullet-proof windscreen.
The first three prototypes built in the DFS workshops had wooden wings and sheet-steel fuselages to which were fitted standard Bf 109 tailplanes. Power was to be provided by two Argus pulse-jets and various engine position configurations were tested in the Messerschmitt wind-tunnel at Augsburg.

In this regard, in July 1941, a series of design projects had been undertaken by *Dipl.-Ing.* Rudolf Seitz, a project engineer in the Messerschmitt *Entwurfsbüro*, together with *Dipl.-Ing.* Prager and *Dipl.-Ing.* Mende, with the aim of producing a small, cheap, expendable, pulse jet-powered, air-launched parasite fighter and bomber plus a mounting, launching and retrieval system. The study was designated P 1079 and was produced in co-operation with the *Deutsches Forschungsinstitut für Segelflug* (DFS) at Ainring. By early 1942, the project had evolved into the P 1079/17 and on 31 March it was submitted to the RLM who assigned it the designation Messerschmitt Me 328. Six versions were proposed; three as variants of the Me 328 A – the fighter variant, and three as variants of the Me 328 B – the fighter-bomber and bomber version.

The Me 328 B-1 was designed to carry a 1,000 kg external bomb load, while the Me 328 B-3 was to have carried a 1,400 kg SD 1400 bomb. It was further planned that these aircraft could be carried by or be towed behind an He 177 or Me 264 by means of a *Deichselschlepp*, or towing pole.[47]

Despite all the development work, if Milch's fantastic plan had ever materialised, then the maximum offensive load which the Me 328 would have been able to carry would have been less than one a half tons – hardly an amount likely to cause significant damage to a major American city or to justify sufficiently the effort and risk to the crews of both aircraft! Nevertheless, the plan would lurk in Milch's mind until mid-1944 (see Chapter Five).

Four days after the meeting with Jeschonnek in Berlin, Milch met once again with *Flugbaumeister* Friebel. During the course of a lengthy discussion, Friebel repeated his opinion that, despite Jeschonnek having rejected the option in February, missions conducted by the Me 264, including the *'Aufgabe*

A dramatic view of the 20 metre long upper surface of the Me 264's starboard wing under completion at Augsburg. The fuel tanks have not yet been installed and on top of the wing is an oil tank. The engineer crouching to line one of the tanks gives an idea of scale. Note the wing attachment brackets either side of the fuselage through which pins have been bolted (see also page 45) to hold the wings and the open chamber for the intended MG 131 Z turret, as well as the circular observation window. Note also that remarkably clear beyond the Me 264 are the typical 'back-up' elements of a production line environment: a sign for a first-aid point, a notice to workers to apply care and a washbasin.

Right The starboard wing of the Me 264 V 1, seen from the trailing edge. The wing has been mated to the fuselage, but awaits the installation of the fuel tanks and control surfaces as well as the fitting of one of its two Jumo 211 engines.

Below Two engineers consult a drawing on the floor of the Augsburg workshop beneath the semi-installed cowling of one of the Me 264's Jumo 211 engines. Note the large blackout curtains hanging over the windows to the right as protection against possible enemy air attacks.

Behälter	kg	Entf. Mitte Rumpf	kg x Entf.	Entf. Hinterk. Fl.	kg x Entf.
1	1140	1940 mm	2 210 000	3920 mm	4 460 000
2	1120	1560 "	1 749 000	2320 "	2 600 000
3	640	3540 "	2 261 000	1400 "	895 000
4	1450	5100 "	7 400 000	3400 "	4 940 000
5	800	9350 "	7 460 000	2700 "	2 160 000
6	1590	5750 "	19 160 000	1920 "	3 270 000
7	1910	8800 "	16 800 000	1620 "	3 085 000
8	405	12750 "	5 170 000	2120 "	860 000
9	685	12750 "	8 730 000	1240 "	850 000
	9740		60 910 000		23 120 000
nutzbarer Inhalt.		60 910 000kgmm / 9 740 kg = 6250 mm		23 120 000kgmm / 9740 kg = 2380 mm	
		Bezugslinie Mitte Rumpf.		Bezugslinie Hinterk. Flügel	

Schwerpunktlinie für vordere und hintere Behälter nach Zeichnung Nr. XV/97.
Entfernung Mitte Hauptspant bis Hinterkante Flügel = 3437 mm.
taer. = 3490 mm/ 0% taer. 10 mm vor Mitte Hauptspant infolge V-Form für Flügel mit 43 m Spannweite.
$3437-2380 = 1057 \times \frac{100}{3490} = 30,4\ \%$

Kraftstoffschwerpunkt auf 30,4 %.

XV/98 Me 264.
10.12.42. Hacke.
veraltet.

10.12.42.

Left A Messerschmitt drawing dated 10 December 1942 showing the fuel tank layout in the port wing of the Me 264. The combined weight of all nine tanks in each wing was 9,740 kg.

Below Using a winch, mechanics gently hoist a fuel tank up towards its bay between the two Jumo 211 engines on the port wing, while an engine technician works on the connections to the outboard engine.

Right The immense bulk of the Me 264 V 1's port wing can be seen from this photograph, which also shows the volume of cabling and wiring at the leading edge. The pressure and load on the wing root must have been considerable. Visible behind the wing is the vertical stabilizer at the rear of the aircraft. The starboard side stabilzer can also be seen immediately adjacent to one of the building's structural girders. Note also the observation window in the fuselage, next to the wing join.

Below Another section of wing leading edge. Note the control rods and the two engine mounts.

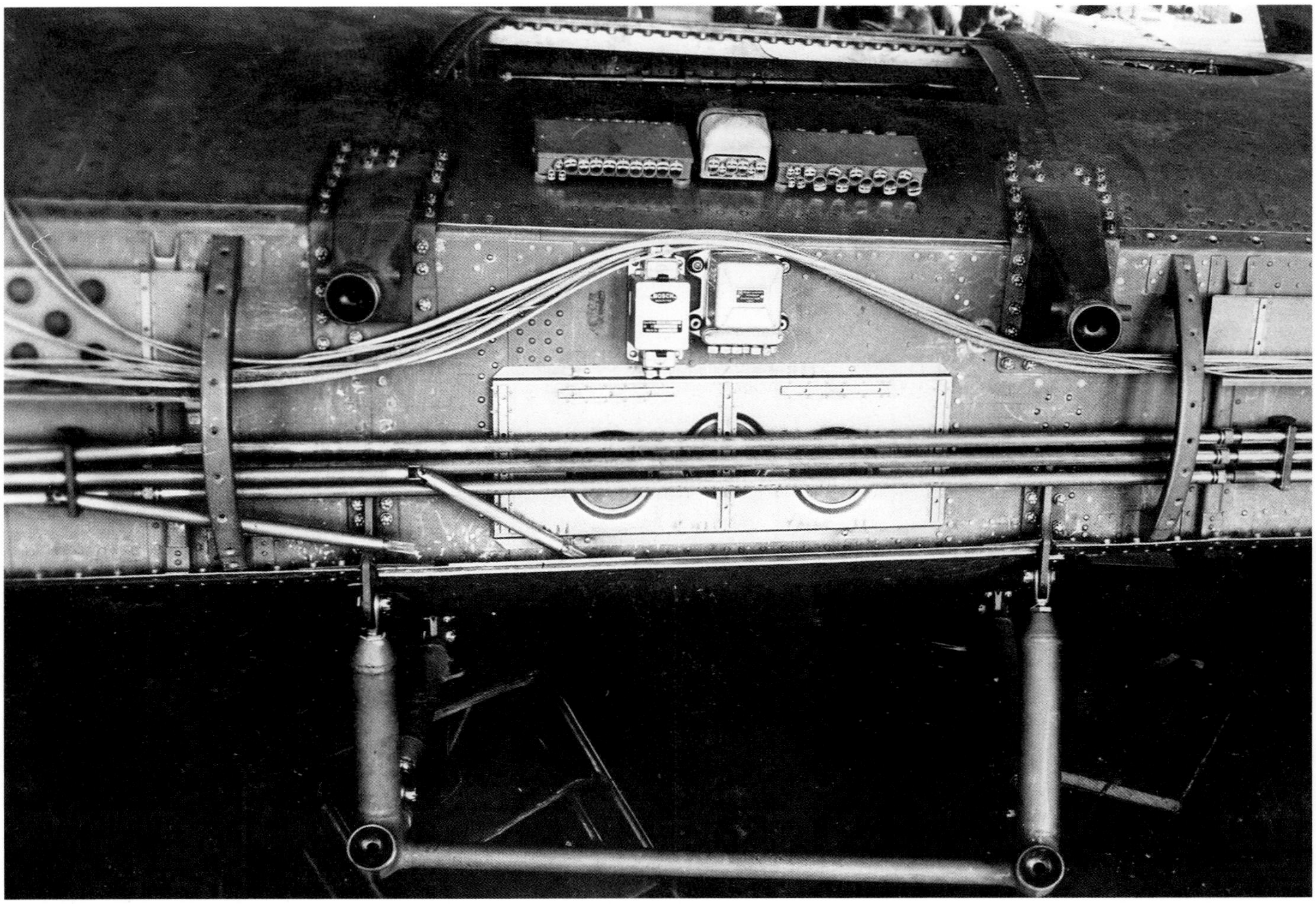

Amerika', which extended over 13,500 km without in-flight refuelling, would not be possible. Weary of taking one step forward and two steps back every time the Me 264 was mentioned, Milch decided that for the time being all further discussions regarding harassment attacks against American targets, and other ultra-long-range reconnaissance missions were to cease.[48]

A view of the almost complete port side vertical stabilizer.

Yet only on 19 May, Milch received more visitors to discuss the Me 264. This time it was Messerschmitt director, Theo Croneiss, accompanied by *Generalmajor* von Gablenz and *Oberst-Ingenieur* Schwencke. Von Gablenz enthused that the aircraft represented an important new step in aeronautical design and development and that it was far in advance, technically, of its competitors. He confidently recommended that work should continue on the Me 264's development and suggested building a run of 30 heavily armed and armoured machines for anti-shipping operations in the 1,500 km stretch of ocean between Newfoundland and New York as well as for long-range harassment missions. This part of the Atlantic was busy with naval and merchant shipping, and the *Luftwaffe* officers reasoned that the Me 264 could cooperate with U-boats. Milch paused to think and then commented: 'There could be a whole host of opportunities there.'

Raids against New York were again brought up. Milch informed Gablenz and Schwencke that distance estimates had been incorrectly calculated. The distance from the French coast to New York had been calculated at 6,700 km while the direct return flight was calculated at 13,400 km, which, including a 10 per cent fuel reserve, amounted to a fuel requirement for 14,750 km. However, as Milch pointed out, Brest was only 5,500 km from New York, the true fuel requirement being for a flight in the region of 12,100 km.

If agreement was given to build 30 Me 264s, they would be able to operate along the entire length of the US Eastern Seaboard striking at key towns and cities, rather than shipping. To dispel any doubts that his visitors may have had as to the value of such a quantity of aircraft committed to such missions, Milch stressed that the prime objective was to tie down US air defences and '... to force the Americans to divert some of their armaments productions to their own defences. We don't have to send a whole air fleet over there. With just a few aircraft much can be achieved. The idea is not to demolish America, but only to force the US to erect anti-aircraft defences. Therefore not only New York but also other areas of the US should be on the receiving end of our bombs. Perhaps we could even fly from Petsamo in Finland over the North Pole to San Francisco. That is probably not much further. Including the 10 per cent reserve it is 7,700 km.'

Theo Croneiss explained that two tons of bombs aboard the Me 264 would shorten its range by 1,200 km, and Milch confirmed that a single SC 1000 reduced the range by 600 km.[49]

These discussions must have shifted Milch toward a more positive attitude on the Me 264 and helped to keep interest in the aircraft very much alive, for according to historian, Manfred Griehl, by early May 1942, the RLM was 'haranguing Messerschmitt for the first 30 Me 264s to be available at the very earliest date'.[50] By early summer, the *Technisches Amt* finally gave its approval for work to continue at both Messerschmitt and Junkers. This was given in the knowledge that the Me 264 represented the most immediate opportunity to strike at the United States, but that consideration should be given to development of the six-engined Junkers Ju 390 which, with an estimated range of 10,000 km, offered far more potential than the four-engined Messerschmitt. Once work on the Ju 390 had progressed to a meaningful point, the possibility existed to use the four-engined Ju 290 and Me 264 as tanker aircraft for in-flight refuelling.[51]

By mid-July 1942 three Me 264 prototypes were undergoing construction at Augsburg. Messerschmitt engineers and technicians projected that the first prototype machine, the Me 264 V 1, would be ready for flight testing on or around 10 October, but again this was optimistic due to delayed and sometimes missing component deliveries. Until matters were firmed up in respect of the V 1, testing dates could not be given for the V 2 and V 3. Production of a fourth prototype was cancelled due to lack of capacity. In his monthly report, Messerschmitt's Production Director, Fritz Hentzen, relates how Messerschmitt was struggling to deal with unreliable suppliers: '*Three prototypes are under construction. Estimated date for the first flight of the V 1: 10 October, V 2 and V 3 unknown. Manufacture of V 4 and subsequent aircraft stopped. All prototype and pre-production versions are suffering from a lack of externally supplied parts and finished components. In particular, the Elektrometall Company at Cannstadt is in serious arrears. Much of the Elektrometall company's employee base has been drafted, and the company was forced to curtail its prototype construction and pass its contract on to subcontractors. The subcontractors in turn had difficulties in meeting demands because some of their workforce have also been called up.*'[52]

Further pressure was coming from Focke-Wulf which had just completed its design of a six-engined 'Long-Range Bomber' project with a maximum range of 13,000 km, powered by either six Jumo 213s, six DB 603s or six Jumo 222s. This 80-ton aircraft featured a fully glazed, pressurized cabin connected by a pressurized corridor to the tailplane with a rear turret housing four MG 151s. Further armament was housed in remotely-controlled side turrets fitted with HD 151Zs. Maximum bomb load constituted three or four racks of SC 2000s. Most of the fuel load was housed in the wings, but auxiliary tanks could also be fitted. The aircraft would also carry an impressive array of radio and direction-finding equipment, as well as two rubber dinghies, 30 litres of water and an emergency transmitter for its crew of six. However, since the required engine types were not available in sufficient numbers and the maximum range was insufficient for a mission to be flown to New York without refuelling, the project was shelved.[53]

Meanwhile, the problems at Augsburg continued. In his report for August, Fritz Hentzen, once again expressed the problems faced by Messerschmitt as a result of supplier difficulties: '*Construction of the V 1, currently in its final stages, cannot be completed in October of this year as previously reported. The undercarriage, which the VDM Company had promised to deliver by 1.9.1942 at the latest, has not yet been supplied and no new delivery date has been set. In general it should be noted that acquiring externally supplied parts and, more recently, even engines, for the prototype aircraft is proving most difficult and is resulting in serious delays in the programme.*'

By late August 1942 a mood of general doubt hung over the entire Me 264 programme. It became increasingly apparent that the chances of the aircraft making its first flight in October grew more unlikely by the day. Messerschmitt attributed this not only to the excessive delays in the delivery of the main landing gear from VDM but also to the engines from Junkers. Once again, the RLM began to look towards the Junkers Ju 290 and the Ju 390 as possible alternatives.[54]

On 7 August, Hermann Göring chaired a meeting at the office of the *Generalluftzeugmeister*. Aside from Milch, the meeting was also attended by Willy Messerschmitt and *Major* Dietrich Peltz. Peltz, a former *Kommodore* of KG 77, was an experienced *Stuka* and bomber pilot, and tactician. He had been awarded the *Eichenlaub* to his *Ritterkreuz* the previous year having flown 250 successful day and night missions. Peltz was visiting Berlin from Italy where he had been most recently serving as commander of the *Verbandsführerschule für Kampfflieger* at Foggia which was in the process of transferring to Tours in France. While in Germany, he had been asked to inspect the Me 264 at Augsburg; he subsequently reported that it had made a very favourable impression upon him. Göring was sceptical: '*You won't find a pilot to fly a crate like that with unprotected fuel tanks for wings.*'

Above In the summer of 1942, Major Dietrich Peltz, a highly experienced Stuka and bomber pilot, was asked to inspect the Messerschmitt Me 264 still under construction at Augsburg. His subsequent report to Milch was favourable. Peltz is seen here later in the war as a Generalmajor and commander of the IX. Fliegerkorps. He is wearing the Knights Cross with Oakleaves and Swords awarded in recognition of his 300 flights against the enemy.

Messerschmitt: '*The* Herr Reichsmarschall *is perhaps confusing it with its predecessor* [Me 261]. *In accordance with the RLM contract I built* this [author's mark] *aircraft with protected tanks, at least in the wings, as the British do, and this aircraft is equipped with the BMW 801 engine for about 15,000 km throttled back.*'

Surprisingly, the *Reichsmarschall* appears to have changed his mind about a 'rushed bombing raid' against the United States by the Me 264 and instead pressed for its use as a mid-ocean convoy spotter for U-boats.

Göring: '*I would hug you if you gave me just a few machines – which don't need to make it to New York – but with which I can make it to mid-Atlantic to get at the convoys...*'

Just over a month later on 13 September, as Messerschmitt was completing its static calculations for the Me 264, Göring was again rueing the fact that only 102 He 177s had been built of which just 33 had been approved for operational service by the *Luftwaffe* Quartermaster-General. Fundamental problems with the Heinkel bomber's coupled Daimler-Benz engines were still being experienced. 'It really is the saddest chapter,' Göring lamented, 'I do not have one single long-range bomber... I look at these four-engined aircraft of the British and Americans with real envy; they are far ahead of us here.'[55]

By late 1942, Edgar Petersen had become a critic of German bomber strategy, for he recognised that the medium bomber, on which so much emphasis, energy and faith had previously been placed, had drawbacks and limitations – including lack of range – and he blamed earlier failures in the air war on the dependence on a medium bomber. At one production conference in 1942, he commented: 'The days of the medium bomber are numbered...'

Thus in the middle of the war, the Germans began to re-examine air strategy, or at least to seek the optimum aircraft and methods of waging a strategic bomber campaign. The problem was made more challenging however, by the immediate demands of the individual war fronts meaning that the technological requirements of a long-range strategic bomber were also hard to meet.[56]

In October, Göring decreed that with immediate effect all new bombers, including those intended for transatlantic operations, were to be fitted with four-gun rear turrets in a similar manner to the British Stirlings, Lancasters and Halifaxes which were bombing Germany. That month Petersen reported to Milch that Messerschmitt was again promoting the concept of a six-engined version of the Me 264. Its performance would far exceed that of the four-engine version. Milch accepted Messerschmitt's optimistic performance forecasts with scepticism; he opined brusquely: '*This Me 264 merely offers propaganda value!*'

For his part, Milch, supported by other senior officers within the RLM, was once again beginning to favour the Ju 390 over the Me 264 as a more realistic contender for a long-range aircraft. The Junkers would have the range; it was expected to be able to fly 10,400 km or 9,200 km with a 3-ton bomb load. Friebel was less optimistic, primarily concerned by the aircraft's low power to weight ratio: 'Well for one thing,' he told Milch, 'the Ju 390 can't get up without help. Towed by an He 111 Zwilling it would need 1,000 metres of runway or, like the Me 264, about 1,300 metres with a four-ton rocket for thrust...'

Milch remained unconvinced; he projected that the Me 264 would not be ready for operations in the *Atlantikkrieg* before 1946/47.[57] However, *Professor* Messerschmitt, refused to recognise difficulties and assured the RLM that on construction of a fourth prototype (the V 4), the company would have completed all work on equipment and defensive armament so that test flights could be combined with 'combat missions against the US in 1944'! From the V 4, the Me 264 would feature a modified cockpit fitted with

Right The port side main undercarriage assembly of the Me 264 V 1 as seen during construction at Augsburg, probably in the second half of 1942. Each wheel measured 1550 mm x 575 mm, and it was planned that each leg would hold a pair of wheels of which one from each pair was jettisonable after take-off. An additional wheel was seen as necessary in view of the fact that the estimated weight of the aircraft was nearing 50 tons. The fixed wheels turned on their axes as they retracted electro-hydraulically.

Left In order to counteract the considerable weight of a fully laden Me 264 on the ground, tests were made using a converted Bf 109 (V 3, W.Nr. 5603, CE+BP) with the idea of developing a twin nosewheel. These tests revealed that compared to a single nosewheel, manoeuvrability had decreased, although there was no shimmying.

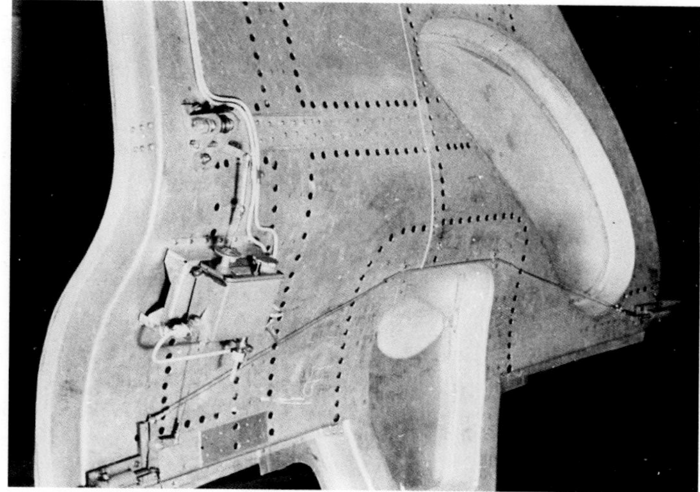

Right The inner surface of the outer landing gear door of the Me 264 V 1, which was positioned between the fuselage and the in-board engine.

a nose turret for harassment attacks against targets on long-range operations. It seems certain departments within the RLM accepted this.[58]

In November 1942, *Oberstleutnant* Theodor Rowehl, the commander of the *Versuchsstelle für Höhenflug* (the Test Centre for High-Altitude Flight) – or VfH – put forward a requirement that all future Me 264s, including the first prototype, should be equipped for high-altitude reconnaissance missions, which essentially meant the inclusion of engine mounts for the new BMW 801 high-altitude engine.

With such redesigns and modifications, weight was becoming a problem. To compensate this, a jettisonable main landing gear was developed, planned for use with the V 2 onwards, which would allow an increase in gross take-off weight from 28,000 kg to 46,000 kg. Messerschmitt proposed developing a twin nosewheel, which was tested using a converted Bf 109 (V 23, W.Nr. 5603, CE+BP). These tests revealed that compared to a single nosewheel, manoeuvrability had decreased, although there was no shimmying. Problems followed with comparison tests between a standard twin nosewheel using 350 x 220 mm tyres and one fitted with large, non-standard 620 x 220 mm tyres, as planned for the Me 264.[59]

On 10 November, the Messerschmitt production team decided that the Me 264V 1 was ready enough to be tested merely as a trial aircraft to assess flight

Below Oberstleutnant Theodor Rowehl, the commander of the Versuschsstelle für Höhenflug. In late 1942, his unit began to take a close interest in the Me 264 as a potential high-altitude reconnaissance aircraft. Although broadly accepting the Messerschmitt design, Rowehl believed the Me 264 would benefit from BMW 801 engines and eventually GM 1 power boost and increased armament. Rowehl had never been a pilot, but during the First World War had served as an Observer with a Marine Feld-Flieger Abteilung. In 1935 he was involved in the establishment of a semi-clandestine reconnaissance unit within Hansa Luftbild, known officially as Flugbereitschaft Abteilung B. In early 1939, this unit was redesignated the Versuschsstelle für Höhenflug which remained in existence until 1944. It later became, successively, the Aufklärungsgruppe Ob.d.L and the Versuchsverband der OKL. A very popular officer, Rowehl was awarded the Ritterkreuz in October 1940 in recognition of his services flying reconnaissance missions.

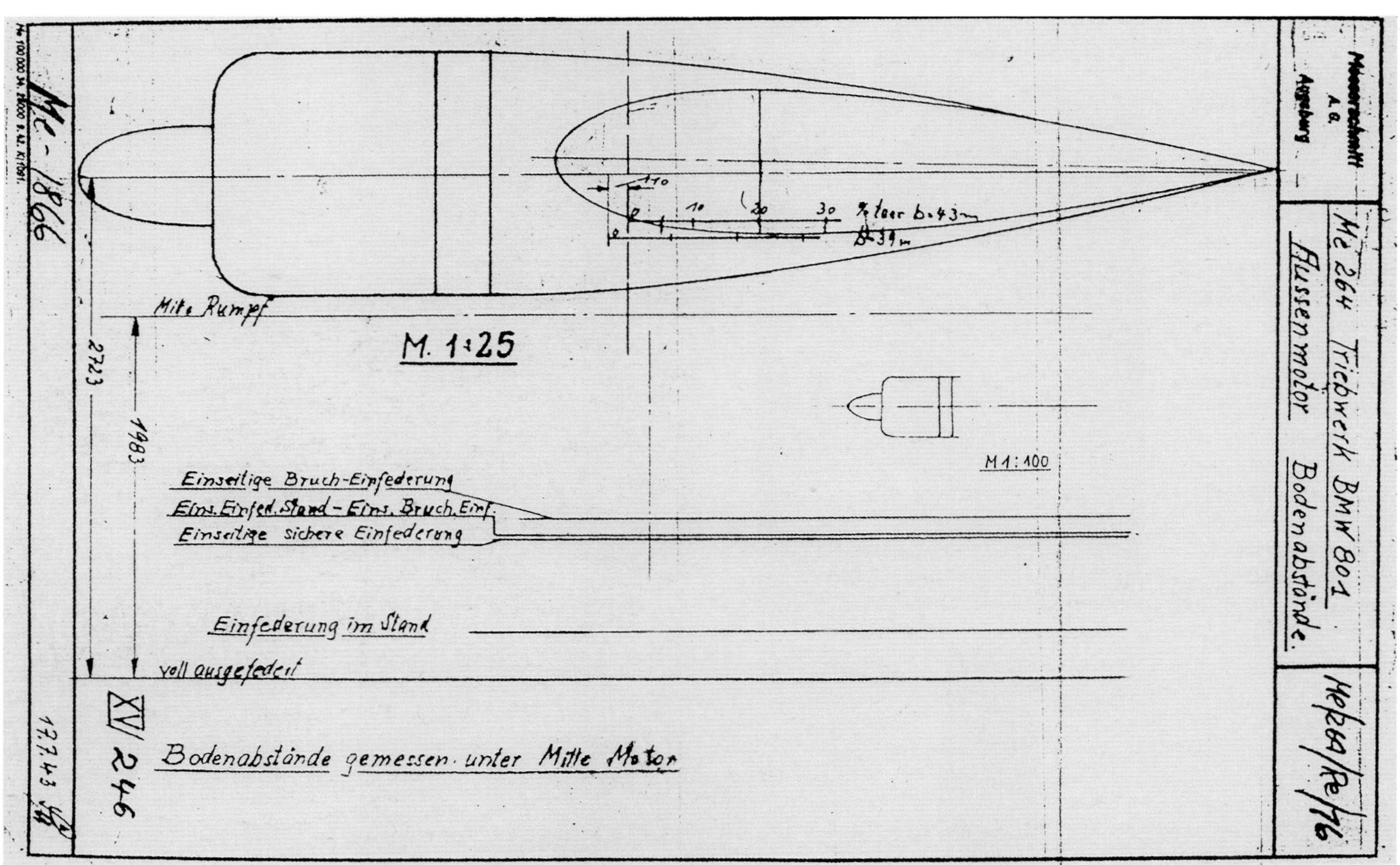

Above A drawing prepared on 17 July 1943 showing the distance from the centreline of the outer BMW 801 engine to the ground – which was 2,723 mm.

characteristics. Fitted with four Jumo 211 J engines, the aircraft – known to the Messerschmitt company under its internal designation as 'M III' – would incorporate a small nose section for the pilot but would not be fitted with armour, armament and gun turrets or bomb release equipment. Simultaneously, Messerschmitt project engineers calculated alternative performances using either four BMW 801 D, or DB 603 H or Jumo 213 A-2 engines. It was felt that a significant improvement in engine performances would be achieved at altitudes up to 8,000 m using the DB 603 H and Jumo 213 compared with the BMW 801 D – but these engines were still not available.[60]

On 11 November, *Dipl.-Ing.* Karl Seifert issued a summary of the latest comparison in performance calculations between estimates produced by the *Erprobungsstelle* Rechlin and by the Messerschmitt project team for a long-range reconnaissance version of the Me 264 fitted with four Jumo 211 engines[61]:

E-Stelle Rechlin	Messerschmitt A.G. calculations	New calculations performance calculations based on weights set by E-Stelle Rechlin	using revised weights as at 2.11.1942
Take-off weight	43,000 kg	43,000 kg	46,000 kg
Landed weight	22,750 kg	22,750 kg	24,700 kg
Take-off weight/ Landed weight ratio	1.89	1.89	1.89
Climbing speed from ground with full operational load	2.35 metres per second	2.5 metres per second	2.0 metres per second
Take-off distance without boost on concrete runway	2,260 metres	2,260 metres	3,030 metres with 47-tons take-off weight
Take-off distance with 4 tons auxiliary thrust on concrete runway	1,150 metres	1,190 metres	1,720 metres with 50.25 tons take-off weight
Maximum speed with full operational load based on 33 tons flying weight at 5.6 km	515 km per hour	520 km per hour	520 km per hour

These figures were to be the last detailed calculations made before the Me 264 finally took to the air for the first time the following month.

The BMW *Einheitstriebwerk* (All-in-one engine unit) – as proposed for the Me 264

Research into optimum performance for the BMW 801 engine

By *Dipl.-Ing.* Erwin Schnetzer, former BMW development engineer

'On 1 December 1936, I was transferred from the Dornier aircraft company to BMW with the specific assignment of converting engines to powerplants by adding cooling and cowling, including oil and exhaust systems as well as engine suspension and automatic temperature control. After initial work on nine cylinder BMW 132-H engines, effort was concentrated on the new 14 cylinder engine, which was in development at this time as the BMW 139.

'The fact that the engine had a powerful cooling fan caused me to abandon the conventional outlet control flaps – '*Spreizklappen*' – and to replace them by a lever-suspended '*Drosselring*', avoiding flow disturbance in climb. Also, the engine intake was taken from the pressure between the cylinders and fan, minimizing external drag.The most critical design item was the oil cooler, which had to be pressure proof to 20 at. to secure throughflow under cold weather conditions. I chose a finned tube Alu system, which could be arranged in the cowling nose, its cooling airflow controlled by a variable nose slot.

The BMW 'Einheitstriebwerk' under test. The engine was proposed for the Me 264.

'For the exhaust system individual pipes were chosen, embedded in closely tailored cowling grooves. Engine and cowling were soft suspended on rubber members. The engine/aircraft interface was standardized, employing quick disconnects for fuel-, oil-, and instrumentation lines. The 'Einheitstriebwerk' prototype was built in-house and flight-tested, after passing a scrutinizing 50 hour test. The aircraft manufacturer then had a choice between buying a naked engine or a complete powerplant.

'The '*Einheitstriebwerk*' was readily accepted for wing installations of two- and four-engined aircraft...

'The BMW 139 engine was conservatively designed based upon experience gained under Pratt & Whitney license agreement, its output being 1,600 hp. Early in 1938, the BMW technical management changed when high-spirited Helmut Sachse, formerly RLM-GL-E3, replaced the old, experienced creator of BMW aircraft engines, Director *Dr.-Ing.*Max Friz. The new era was marked by new engine designations and the tendency to deviate from Pratt & Whitney practice was scuppered and replaced by the 802 engine which incorporated radical changes preventing it from ever passing a 50 hour test.The changes from 139 to 801 were less drastic and consisted of design refinements, such as increased cylinder finning, 2-speed compressor and – most importantly – the introduction of single lever control, called '*Kommandogerät*'. The external dimensions of 139 and 801 were identical and the cowling, oil cooler, exhaust and cooling air control concept remained the same. The engine installation design was refined in many details, such as incorporation of temperature sensingVDM hydraulic control units for engine and oil cooler air sided control, also the quick disconnect firewall interface was perfected.'

Mindful of the duplication of effort in developing both the similar BMW 139 and Bramo 329 engines the RLM decided, on 8 June 1939, to combine the companies under BMW's stewardship, Bramo then being known as 'BMW Flugmotorenwerke Brandenburg GmbH'. At the same time Helmut Sachse, formerly with the RLM GL/C E3 department, took over from BMW's director, Max Friz. The reorganisation was also marked by a change in engine designations, work on both units being terminated in favour of the BMW 801 project. A total of 47 BMW 139s were built before the Munich factory switched to the new BMW 801.

From then on the number '1' for BMW engines and '3' for Bramo engines was dropped in favour of '8' for the combined company.

Engine designations

Real engines under development or in production were designated with 3 digit model numbers again starting at BMW with '8' followed by capital letters to identify different versions of each basic engine model. Example: BMW 801 C. Dash numbers were added to differentiate between successive production runs with minor differences. Example BMW 801 C–2

There were three basic ways to buy an engine:

a) 'Nackter Motor' (naked engine), e.g. BMW 801 A, the complete engine with all accessories, baffles, cooling fan, mounting ring and propeller pitch changing mechanism but without cowling, oil cooler and exhaust system.
b) 'Motor Anlage' (powerplant), e.g. BMW 801 MA, naked engine plus cowling with oil cooler but without exhaust system and mounting structure.
c) 'Triebwerk-Anlage' (propulsion system), e.g. BMW 801 TA, the complete propulsion system consisting of 'Motor Anlage' plus exhaust system and mounting structure ready to be attached to the firewall of the airframe.

Chapter Four

'Sudeten'

'A beautiful aeroplane...'

***Flugkapitän Dipl.-Ing.* Karl Baur, 23 December 1942**

On the cold, winter day of 23 December 1942, the Messerschmitt Me 264, V 1, W.Nr. 26400001, coded RE+EN, was rolled out of its hangar at Augsburg ready for its maiden flight. The aircraft, weighing 21,175 kg, was fitted with four 12-cylinder, liquid-cooled Junkers Jumo 211J-1 engines, the same as those used on the Junkers Ju 88 A-4, together with Ju 88 nacelles and radiators, but carried no armament or gun turrets. The flight-test programme was under the overall control of *Dipl.-Ing.* Gerhard Caroli and the pilot was *Flugkapitän Dipl.-Ing.* Karl Baur, one of Messerschmitt's most experienced test-pilots who had been closely involved with the Me 264 since construction of the V 1 prototype had first started.

The completed Messerschmitt Me 264, V 1, W.Nr. 26400001, coded RE+EN, emerged from its hangar at Augsburg for the first time on 23 December 1942. It was fitted with four 12-cylinder, liquid-cooled Junkers Jumo 211J-1 engines but carried no armament.

As a first stage, the aircraft underwent an extensive taxiing test, before moving to the start line on the runway. Take-off was trouble-free, although because of some safety concerns, the landing gear was left down for the duration of the flight. Baur reported:

'*Take-off presented no particular problems. Holding things in a straight line was not a problem. In fact, the nosewheel could have been retracted halfway through the take-off run. The angle of the aircraft was 3 degrees on take-off and was such that the nosewheel left the ground and could have been retracted.*

'The rudder pressure was too high, but by using the trim tabs this problem was overcome... The aileron controls are quite sharp and can be compared with the Me 261 V 3 which has a 90 sq m wing area.

'There was a sudden jolt when dropping the landing flaps, which was compensated by adjusting the elevators.

'The green indicator light to show that the landing flaps are extended did not function. Following landing, and having carried out an inspection, it was found that one of the hinge points was bent which prevented the flaps from fully extending.

Above Dipl.-Ing. Gerhard Caroli, the head of flight testing at Messerschmitt A.G., is seen third from left in this photograph taken at Lechfeld in the summer of 1944, between the shoulders of Hauptmann Werner Thierfelder, the commander of the Me 262 test unit, Erprobungskommando 262, and Gerd Lindner, an Me 262 test pilot. Caroli was placed in charge of the Me 264 test programme and as such was instrumental – along with Karl Baur who can be seen kneeling to the bottom right of the photograph – in the troubleshooting and day-to-day technical development of the aircraft.

Inclement weather forced Baur to end the flight prematurely after 22 minutes. However, on landing the aircraft needed the entire length of the runway due to a brake system failure and rolled off the end into a ploughed field.[1] In his Flight Report, Baur recorded the following problem areas:

The brakes did not function
The landing flaps and suspension were damaged
The aileron indicator did not function
The landing flaps were to be as per the Me 261 and have instruments which indicated the flap positions
Some of the instruments in the cockpit needed to be repositioned such as the rev counter indicator and the pressure gauge which Baur asked to be moved to within the pilot's field of vision.

However, he later recorded of the Me 264's maiden flight:

'A beautiful aeroplane. It was equal to the American B-24 bomber. Only minor problems occurred during the first flight. The worst, the brakes of the landing gear did not work properly. I managed to stop that large aircraft in a field adjacent to the end of the runway.'[2]

As the flight-testing progressed, over the next eleven such flights up to 6 March 1943, Baur would report a number of re-occurring problems which were to dog the aircraft from the onset in this early phase of its development. For example, the forces on the controls were far too high, particularly those related to the tailplane, and Baur also complained about exhaust fumes penetrating into the cockpit.[3]

As the Me 264 embarked upon its first flight-tests, so Junkers had, in the meantime, begun work on its six-engined Ju 390. It seemed to many that this aircraft was better suited for attacks on the USA, and therefore Messerschmitt was forced to make amendments and design enhancements to his own aircraft for use on extended-range maritime reconnaissance missions.[4] However, the *Kriegsmarine* did not favour the Messerschmitt bomber for reconnaissance work, preferring instead the planned Ta 400, although this aircraft was not expected to be available before 1946. Conversely, the *Luftwaffe's* Ordnance Department told Messerschmitt '...to hurry the completion of the two remaining Me 264s for so-called 'Atlantic Missions'.'[5] One serious problem however, was that no aircraft facility existed which was capable of producing up to 30 Me 264s as had been called for.

Meanwhile, after a string of triumphs in the East, the war was beginning to turn against Germany. In North Africa in early November 1942, British forces under General Montgomery were driving back Axis forces from their advance into Egypt, while on the 8th, the Allies launched Operation Torch and landed 107,000 men in French North Africa. In the East, the Russians would not let go of Stalingrad, winter was coming, and yet aircraft were urgently needed in the Mediterranean. In Germany, the inventive mind of the former *Kommodore* of KG 30 and *Ritterkreuzträger*, *Major* Werner Baumbach, continued to work; as he recalls in his memoirs: '*After the Anglo-American invasion of North Africa I had proposed that we should inaugurate small strategic raids, which would have great effect, on North America while we were still in possession of advanced bases*

Below The low profile and extremely low ground clearance of the Me 264 is evident from this view of V 1, W.Nr. 26400001, RE+EN, seen at Augsburg or Lechfeld in late 1942 or early 1943. Note also the thickness of the wing leading edge and the size of the main gear wheel wells, apparent from the space between the outer and inner undercarriage doors.

Karl Baur

Perhaps more than any other man, Karl Baur was deeply involved in the flight-testing and aeronautical assessment of the Messerschmitt Me 264. He flew the '*Sudeten*' on at least thirty occasions and his subsequent reports contained information that was vital to the Messerschmitt designers and project engineers.

Karl Baur was born in Laichingen, Württemberg, in south-west Germany on 13 November 1911, the fourth child of restaurant and butcher shop owners, Jakob and Katherine Baur. His childhood was happy, but he left home at a young age in 1923 to attend school in Stuttgart. In the summer of 1929, Karl enrolled on a four-week summer youth camp at Böblingen for those interested in the possibility of learning to fly. The camp had been organised by veteran pilots of the First World War, and Karl's first introduction to an aeroplane was the Heinkel He 21 biplane. He never looked back and flying became his life. Initially, he satisfied his desire to fly by building model gliders and won a long distance competition when his rubber band-powered model managed to fly 100 metres!

For a while, Karl lived with relatives in Stuttgart while he worked in local motor car and aircraft workshops (the latter being one of the few permitted to build small, single-engine sports aircraft under the terms of the Treaty of Versailles) as part of his 12-month apprenticeship.

During the late 1920s/early 1930s, Karl spent as much time as he could on the Wasserkuppe mountain learning to fly gliders along with hundreds of other young Germans. Here he got to know Wolf Hirth, a famous pre-Nazi period glider pilot and one of the great pioneers of post-war German aviation. Baur impressed Hirth and Hirth invited him to take part in events in which his *Akaflieg Stuttgart* (one of a number of flying clubs established for students and academics) took part. According to Isolde Baur's biography of her husband: '*Karl would fly any machine that was offered to him.*'

Flugkapitän Dipl.-Ing. Karl Baur (right) enjoys a cigarette in the company of a member of his groundcrew in front of the glazed nose of the Me 264 V 1 at Augsburg in 1943.

By the summer of 1933, the year in which Hitler came to power, the *Akaflieg Stuttgart* had voted Karl to be its President. Baur felt uneasy about the Nazis and endeavoured to resist initial attempts by them to roll up the *Akafliege* into the newly-formed National Socialist Flying Corps – the NSFK. However, Baur continued his association with Wolf Hirth and also *Professor* Walter Georgii, head of the *Deutsches Forschungsinstitut für Segelflug* (DFS), working with them as an instructor at Hirth's soaring school on the Hornberg mountain at weekends. In March the following year, he undertook a 6.5 hour duration flight over a snow-covered Hornberg while flying an *Akaflieg*-designed F1 *Fliedermaus* sailplane following which he was awarded his International Silver 'C' Badge.

In September 1935, Baur travelled with Wolf Hirth overland, via Siberia, to Japan where there was a requirement for two qualified German soaring instructors. In Japan, Hirth and Baur were given honorary Japanese military ranks, and offered their services to train glider pilots at Tokorogawa and Ueda, where they used their own sailplanes and a Klemm tow plane to instruct their pupils.

Returning to Stuttgart in February 1936, he applied for his Masters Degree in Engineering from the University of Technology in May of that year and then joined the *Deutsche Versuchsanstalt für Luftfahrt* (the German Test Establishment for Aviation) – the DVL – at Berlin-Adlershof in November. Here he was assigned the task of co-ordinating the various activities of the *Akafliegs* within the universities, including organising trans-Alpine gliding flights. Between July and October 1937, at the request of both German and Portuguese Governments, Baur worked as an instructor assigned to the first soaring school in Portugal. Using his natural talent for languages, he conducted his lectures and seminars in French.

Following brief spells at the DVL in Berlin and a *Luftwaffe Fliegerschule* in Sorau in 1938, Baur was then assigned to the *Erprobungsstelle* Rechlin where he undertook an investigation into how trained engineers could be used as first-class test pilots for the aircraft industry. Baur was shocked at how little had been done in this area and tried to co-ordinate the training of pilots with engineering degrees, but it was not an easy task; there was an astonishing quantity of aeroplanes available to him, but such was their age and variety – including pre-First World War designs and abandoned prototypes – that it was virtually impossible to introduce any firm measures. As he wrote later: '*We worked closely together with the* Luftwaffe Erprobungsstelle *at Rechlin. However desperately needed items were not available to us, such as instruments, special testing equipment and other supplies. We were short of everything from paper to pencils. So we never got beyond the start point...*'

By August 1939, Karl Baur had flown 87,000 km, in 1,121 hours over 2,855 flights in some 60 different types of aircraft.

Then in March 1940, he was offered the position of test pilot with Messerschmitt A.G. in Augsburg, as a replacement for *Dr.-Ing.* Hermann Wurster, the Chief Test Pilot for experimental aircraft and the man who had won the first speed record for Germany when he flew at 611 km/h in a Bf 109.

Throughout a long and accomplished career with Messerschmitt, Karl Baur flew the Bf 109, Me 209, Me 309, Me 210, Me 410, Me 261, Me 264, Me 321, Me 323, Me 163 and Me 262. On 30 April 1943, he was awarded the *Kriegsverdienstkreuz 1. Klasse* (War Service Cross, 1st Class).

Shortly after the cessation of hostilities in May 1945, Baur, together with his immediate superior and colleague from the Messerschmitt testing department at Augsburg, Gerhard Caroli, worked with Colonel Harold E. Watson of the USAAF and his unit, 'Watson's Whizzers', to demonstrate key German aircraft for American evaluation as part of 'Operation Lusty'. On 30 May 1945, for example, Baur took Colonel Watson on a 16 minute familiarisation flight in a two-seat Me 262 B-1a jet trainer. He also assisted Watson in flying an Arado Ar 234 from Stavanger in Norway to France in readiness for onward shipment to the US. Nevertheless, while the Americans found Baur's contribution to their activities professional, it was more restrained than many of his other former colleagues now working for the US forces and American Military Intelligence.

Above Karl Baur conducts an instrument check in the Messerschmitt Me 262, the world's first operational jet fighter. Baur is sitting in Me 262 V5, W.Nr. 130167, an aircraft which was used to test a variety of equipment including the EZ 42 'Adler' (Eagle) gun sight, seen above the instrument panel directly in front of him. The sight automatically calculated the angle of deflection required to hit a target when both attacking and target aircraft were manoeuvring. It was often not reliable.

Above Karl Baur prepares to lift open the access panel to inspect the so-called 'Lotfe Kanzel' glazed bomb-aimer's compartment in the Me 262 A-2a/U2, W.Nr.110555. It was intended that the bomb-aimer would lie prone in this compartment to operate a Lotfe 7H3 bombsight. Trials with ETC 504 bomb racks were conducted in late 1944, but no further progress was made with the idea. On 30 March 1945, the aircraft was used by a defecting German pilot to fly to the Allies.

After subsequently working for a brief time for the Americans at Wright-Patterson AFB in the United States, Baur returned to Germany and his wife in December 1945. In 1948, he took a job as a sales representative with a car battery manufacturer in Stuttgart. Then, in August 1954, he was offered a position with the Chance Vought Aircraft Corporation in Dallas. As his wife Isolde has recorded: '*Financially bankrupt and in poor health he arrived with his family in Dallas hoping to rebuild his life, enjoying the easy-going atmosphere of this great country. Karl Baur lived up to his commitment, fighting all odds that came along, never counting on financial success, but always ready to teach and counsel his fellow pilots, no matter in which country they lived or which language they spoke.*'

Flugkapitän Dipl.-Ing. Karl Baur died in Texas on 12 October 1963.

in Normandy and on the French Atlantic coast. I developed that idea at a private meeting with Jeschonnek... I thought it would certainly have a great effect upon morale if we carried out a "retribution" raid on New York. In this connection I made another proposal which was feasible with aircraft and crews then available. It was a surprise attack on the American continent which we thought would create real consternation among the civil population and provoke panic measures in the way of air-raid precautions, air defence and so forth. A few flights would do great harm to the American economic and industrial programme as large quantities of personnel and material were held up in the USA. For hours Jeschonnek forgot his worries about the bogged-down air war in the East... He sat up with me discussing all the details of my scheme until long after midnight.'

But Baumbach then wrote that the idea was '*...soon forgotten.*'

Meanwhile, the Italians were still trying to devise plans to attack the elusive target of New York. In the Italian Air Ministry building on 7 February 1943, *Generale* Eraldo Irali of the *Regia Aeronautica*, chaired a meeting at which the possibility was discussed of using the four-engined Cant Z.511 floatplane to conduct a bombing mission against the American city. This latest plan had been put together by the qualified team of the aeronautical engineer, *Sottotenente* Armando Palanca, who had been involved in planning a long-range flight from Italy to Tokyo the previous year and who was a production inspector for Alfa Romeo, FIAT and Piaggio, and *Capitano* Publio Magini who, as a pilot, had taken part in the Japan flight.

Left The Cant Z.511 floatplane speeds over the water during float tests off the Italian coast. In February 1943, the Regia Aeronautica devised a plan to attack New York using the aircraft which was to be refuelled in mid-Atlantic.

This page The Me 264 V 1, W.Nr. 26400001, RE+EN was completed in a standard RLM 70/71 splinter pattern, with RLM 65 undersides. From the cockpit, the demarcation line rose from the lower fuselage up to the wing roots and then back down to the lower fuselage, extending to the tail assembly. The fuselage code was applied in prominent black lettering. As far as can be seen from known photographs, no Werknummer was applied to the aircraft.

Above A rare view of the port side of the Me 264 V 1, W.Nr. 26400001, RE+EN seen at Augsburg. Note the fuselage observation window between the code letters 'R' and 'E'.

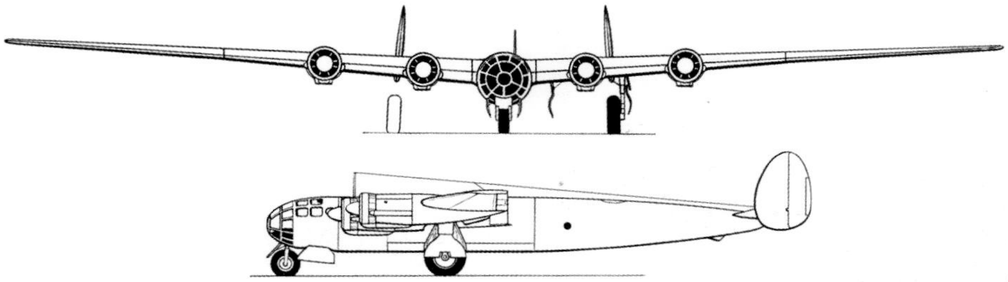

With four Jumo 211 engines

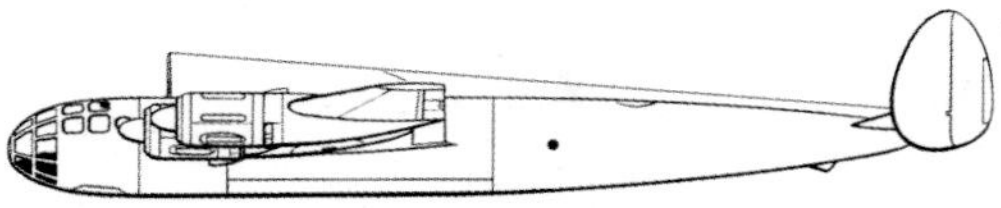

With four BMW 801 engines

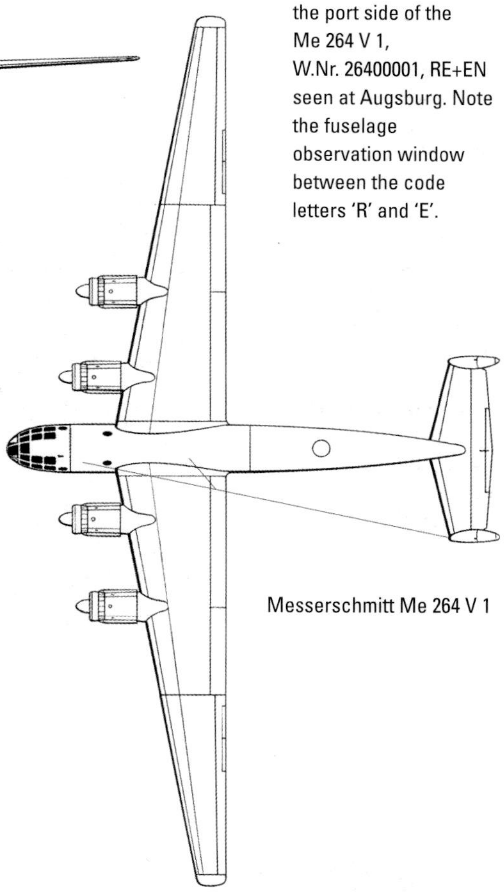

Messerschmitt Me 264 V 1

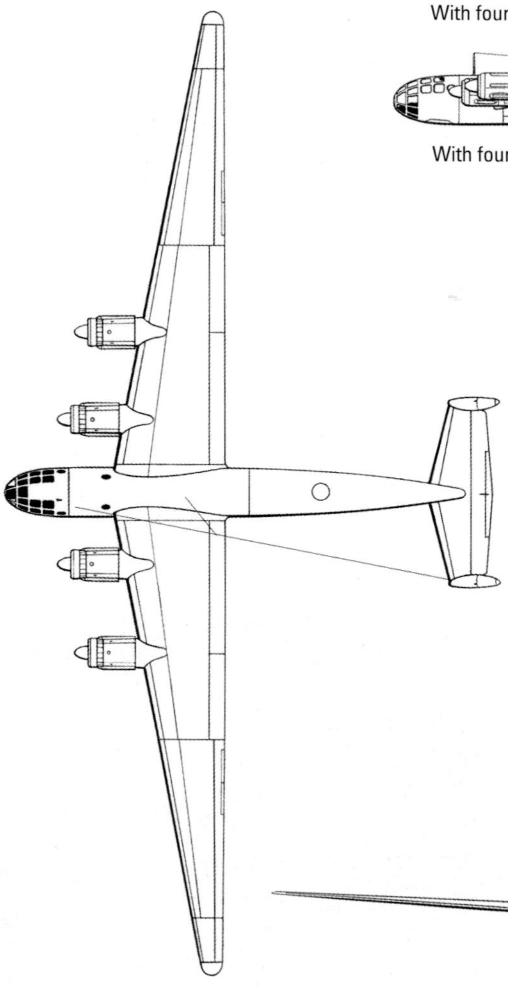

Proposed Messerschmitt Me 264 V 1 with long span wing

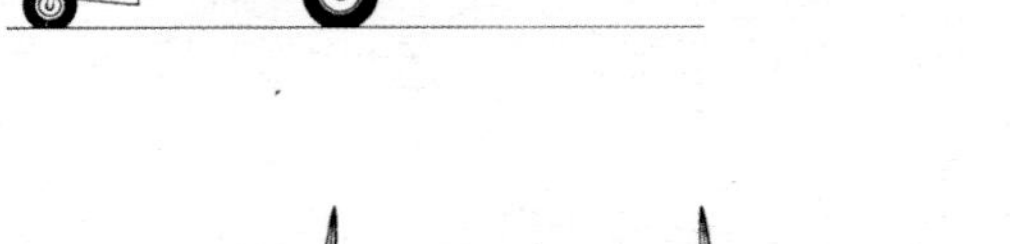

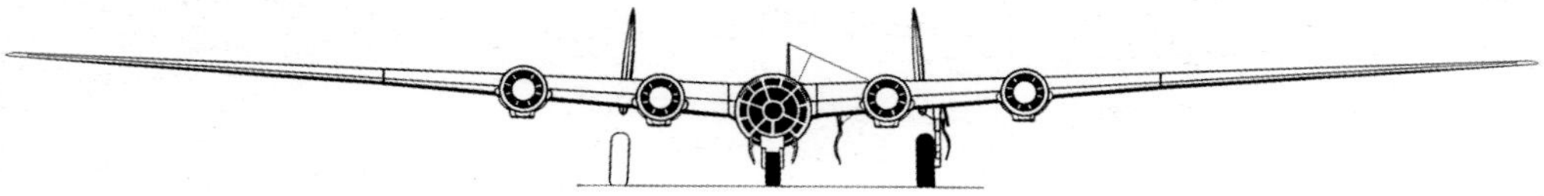

Above The virtually completed cockpit of the Me 264 V 1. The leather-backed pilot's seat in the photograph of the mock-up in the previous chapter has been replaced by an armoured one, but the co-pilot's seat has not yet been installed on its runners. Note also the absence of the 13 mm MG 131 machine gun as seen in the mock-up. The main instrument panel directly ahead of the seats contains the compass. The central console between the seats holds the secondary compass, control box for RATO units, propeller controls, oil temperature gauges, FuG 16 and indicator lights. Note the oxygen unit on the fuselage wall to the right of the co-pilot's position. Note also the lifting handle on the metal unit immediately behind the control console to allow easy access to wiring. The cockpit glazing is heavily smeared, probably with condensation and general grease associated with the environment in the workshop.

Above Close-up of two oxygen supply units in the fuselage compartment. These would probably have been for the flight engineer and gunners. Note a similar unit on the fuselage wall in the photograph at the top of the page.

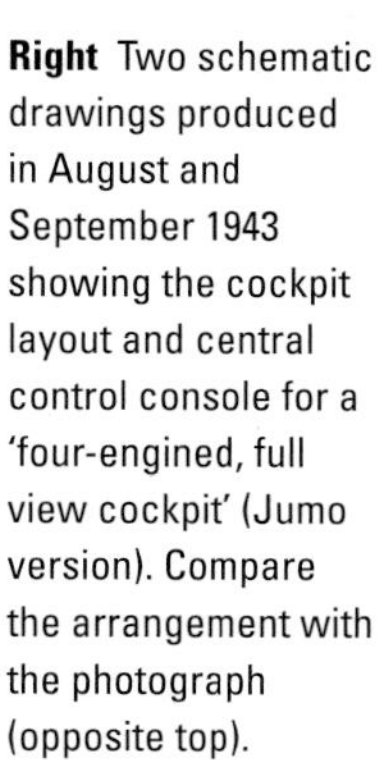

Right Two schematic drawings produced in August and September 1943 showing the cockpit layout and central control console for a 'four-engined, full view cockpit' (Jumo version). Compare the arrangement with the photograph (opposite top).

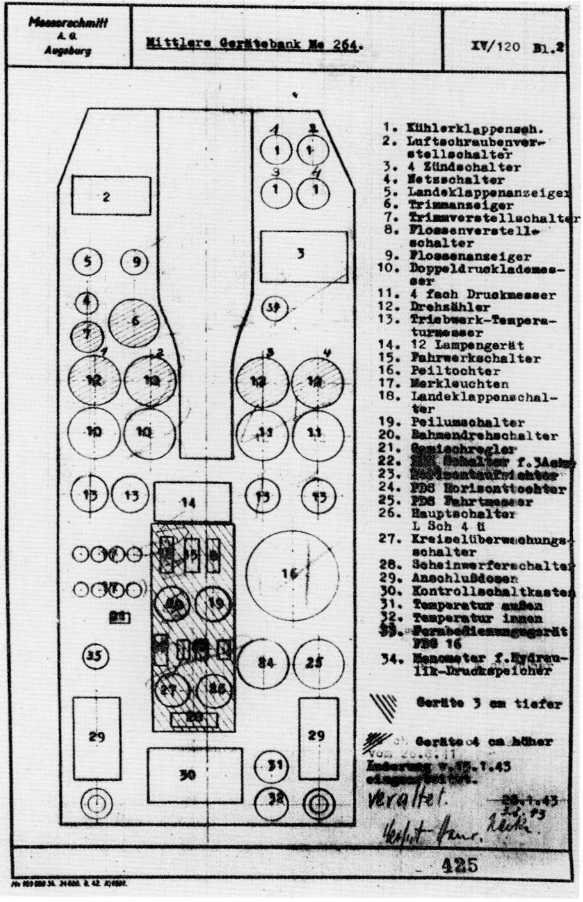

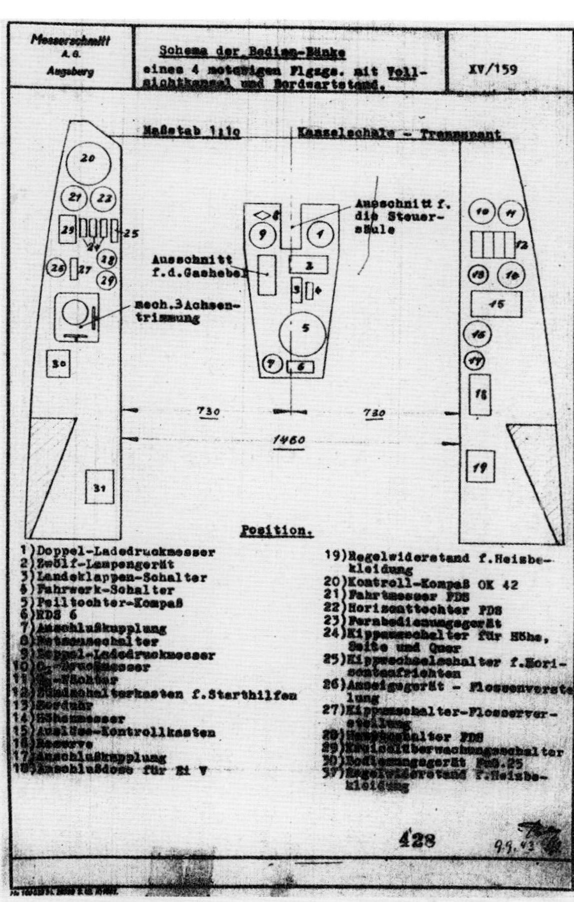

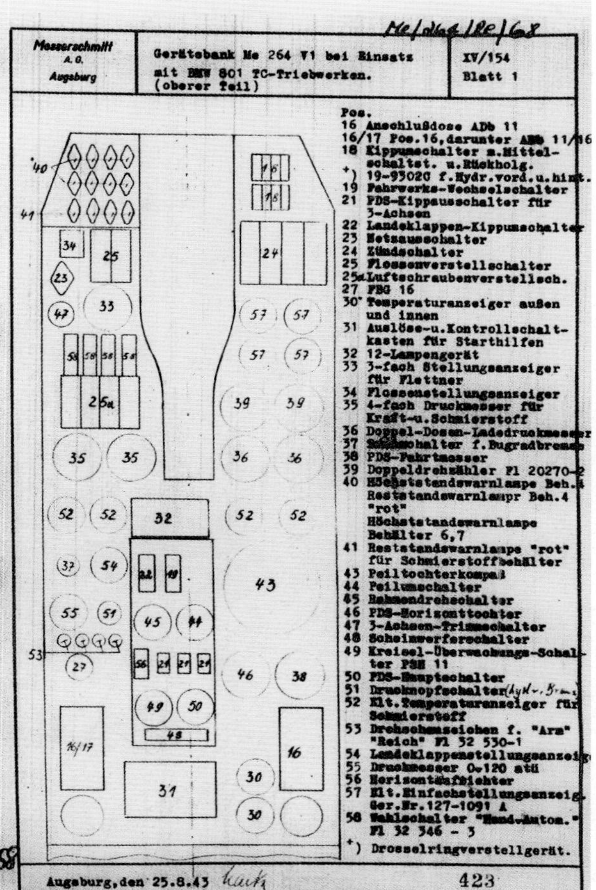

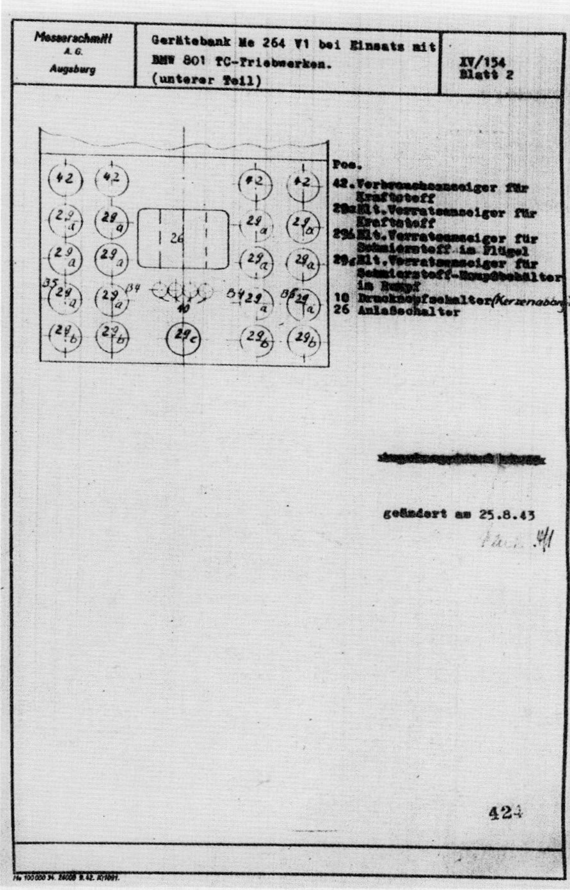

Left Two revised drawings dated 25 August 1943 of the instrument console of the Me 264 V 1 fitted with BMW 801 TC engines.

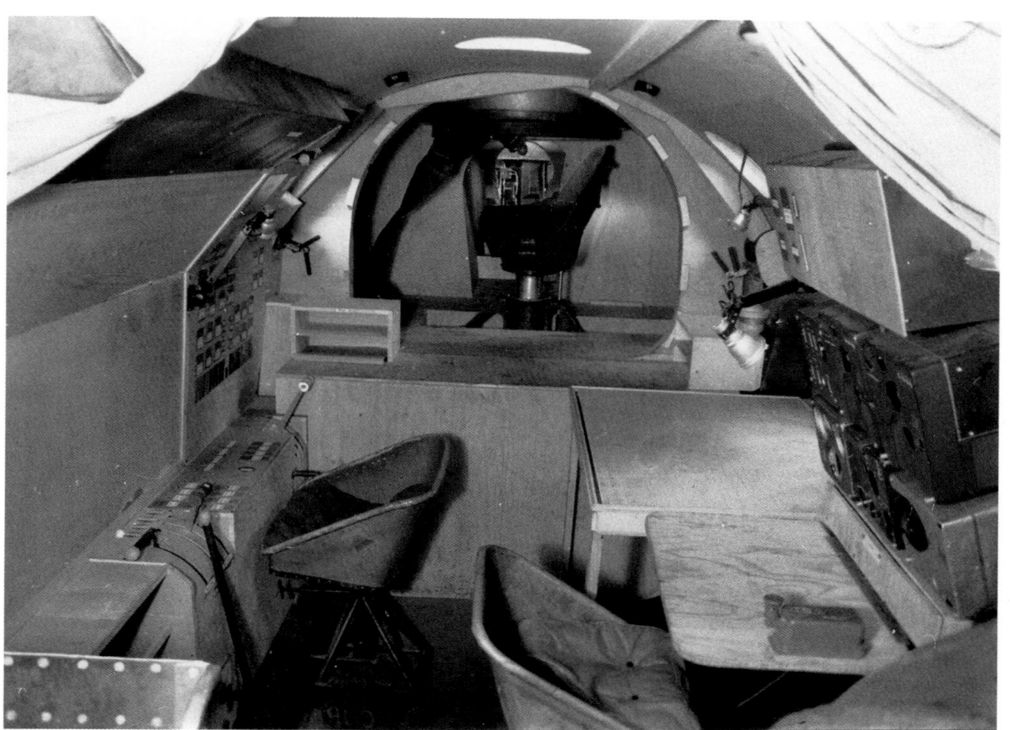

Left The view looking back through the fuselage from the cockpit. To the right is the navigator and radio operator's station with fold-down table and a larger, fixed map table beyond. Two wall-mounted anglepoise lamps provide illumination, but natural light would also have come from the observation windows directly above and in the side of the fuselage just before the access to the central fuselage. On the other side of the fuselage is a control panel used by the flight engineer. The access arch behind these stations leads through the fuselage, above the bomb bay and past a gun position seat. The seats seen are unlikely to have been the final models, which leads to the conclusion that access space for aircrew with the bulk of flying suits and oxygen equipment would have been quite restricted.

Below The Me 264 contained an astonishing mass of wiring, cable and ducting as can be seen from this photograph of a section of the port side fuselage wall. The wording on the metal junction boxes reads 'Schutzkappe erst beim Einsetzen des Gerätes entfernen!' – 'Close lid first before activating equipment!'. Notice the crew entrance/exit hatch to bottom left.

Far left View looking towards the rear of the aircraft, through an access arch, aft of the bomb bay. The circular 'dish' in the roof of the fuselage is probably a compass.

Left A fuselage bulkhead from the tail end of the aircraft, with attachment points, valves and electrical servos.

Palanca had been asked by the Chief of the Air Staff, *Generale* Rino Corso Fougier, to assess the technical feasibility of a New York operation. The Z.511, which was at the time at Vigna de Valle having undergone fuel consumption tests in preparation for long-range flights, was considered of sufficient performance to be able to attempt the mission. However, this could only be achieved if the aircraft used the French port of Bordeaux as a starting point and would be able to refuel in mid-Atlantic from a waiting submarine tanker. Furthermore, there was concern at the potential impact from the drag caused by its large floats as well as possible problems from its Alfa 135 engines. These factors, combined with the risk of a water landing in the Atlantic under radio silence, were deemed to be too great and the idea was dropped.

The day after the meeting at the Ministry, 8 February, Palanca, Magini and a small team, travelled to Vergiate to meet with technicians and engineers from the SIAI and Salmoiraghi firms to discuss refinements to be made to the four-engine SM.95, another aircraft put forward as a possible machine able to strike at New York. Palanca had calculated that, equipped with contra-rotating Alfa 128 engines and at optimum altitude with engines set for minimum fuel consumption, range could be stretched to 11,000 km, allowing a bomb load of just two 250 kg bombs – hardly enough to inflict significant damage, but enough to alarm the world. The plan to use the SM.95 against New York underwent a number of revisions, including one idea to 'bomb' Manhattan with thousands of propaganda leaflets in an attempt to raise the large Italian-American population there against the US Government. Ultimately, however, all the Italian plans came to nought.[6]

Left Having shelved plans to use the Cant Z.511 to strike at New York, the Regia Aeronautica then considered using the SM 95 – seen here – as a 'leaflet bomber' over the city. The aircraft is seen here nearing completion at Vergiate in the spring of 1943.

On 20 January 1943, Karl Baur was at the controls of 'RE+EN' once again when he took it into the air for its second flight from a rain-wet Augsburg. Apart from struggling with the poor layout of the controls and the smell of fumes which filled the cockpit, he found the rudder and aileron forces still too high and they needed to be adjusted by the trim tabs. On landing, Baur encountered the same problem as he had done on the first flight, when again the aircraft needed the entire length of the runway. However he was able to bring the aircraft to a stop just in time. Baur attributed this to the fact that the underside of the aircraft had become covered with wet mud and dirt from the ground which had affected take-off. It had become obvious however, that on a dry runway, the required landing distance would have been much less. Baur also reported that, similarly, during take-off the glazed nose became spattered with mud and dirt from the nosewheel. Rudder pressure had to be reduced by using the trim tabs. Elevators appeared to function without problem. Finally, Baur noted that the nosewheel felt unbalanced and the landing flap on the starboard, in-board wing was damaged.[7]

It was eventually decided to transfer the testing 20 km further south to Lechfeld, near Schwabmünchen, where there was a sufficiently long concrete runway, but this was delayed due to the bad weather. Although Lechfeld had only one hangar large enough to accommodate the Me 264, conditions there were better than at Augsburg where the soft, grassy surfaces presented problems for such a large, heavy four-engined aircraft.[8] However, the wet weather foiled the transfer and on the fifth flight from Augsburg, the underside of the fuselage was slightly damaged when it accidentally scraped the ground. The hydraulics system also failed, resulting in the landing gear becoming jammed and unable to retract.

On 27 January, *Dipl.-Ing.* Karl Seifert held discussions with representatives of *Oberstleutnant* Theodor Rowehl's *Versuchsstelle für Höhenflüge* in order to draw up a mutually acceptable specification for a revised

Below and bottom The Me 264 V 1's outer 12-cylinder Jumo 211 engines are run up at Augsburg or Lechfeld in late 1942 or 1943.

This page The Me 264 V 1 in flight. At least two of these photographs may well have been taken during the aircraft's maiden flight over southern Germany on 23 December 1942 when because of safety concerns, the landing gear was left down.

Fernaufklärer version of the Me 264 which would also have assault transport capability and be able to accommodate paratroops. It was agreed that four air-cooled BMW 801 G engines should be used, since these were at a more advanced stage of development than the DB 603 and Jumo 213 alternatives, and they would include a GM 1 power boost facility of 600 kg thrust for 25 minutes. The aircraft would carry three RB 50/30 cameras, but would feature no armour protection, pressure cabin, or heating or de-icing systems. Armament would consist of one MG 131 in the '*A-Stand*', one MG 131 in the '*B-1 Stand*', one remotely-controlled MG 151 in an HD 151 hydraulic rotary ring mounting in the '*B-2 Stand*', one MG 151 in the '*C-Stand*' and one MG 151 in retractable mountings in each of the two '*D-Stand*' 'waist' positions.

Range was estimated at 15,500 km with a take-off weight of 56 tons, or 13,000 km at a take-off weight of 44.2 tons without external fuel tanks. For such take-off loads, Seifert included a jettisonable undercarriage in the design specification, in the form of two additional tandem wheels fitted to each fixed wheel. The aircraft would require a runway length of 1,900 metres without RATO units, or 1,100 metres with RATO units fitted. Once airborne and with auxiliary tanks jettisoned, speed was expected to be 510 km/h at 6,500 metres. The aircraft would have its bomb-bay area removed so as to accommodate space for a team of 20 seated paratroops or soldiers in a rearward compartment together with an exit ramp. Quite how it was foreseen to use 20 lightly-armed parachutists effectively on such long-range missions is not clear.[9] Nevertheless, a detailed 40-page *Baubeschreibung* for an '*Me 264 Fernaufklärer*' was issued by Messerschmitt at Augsburg on 15 March 1943. This document was later obtained by the British Ministry of Supply which made a translation (see Appendix Two).

Meanwhile, the Me 264 V 1, by now codenamed '*Sudeten*', was transferred to Lechfeld on 22 January 1943. Flight-testing continued throughout February and March. On 1 February, Karl Baur was airborne on the sixth flight to determine certain flight characteristics and to test persisting problems with the retraction of the landing gear. He discovered that the gear still did not retract properly. Furthermore, although the elevators responded easily, the aileron and rudder forces were still found to be excessive at high speeds and changes made had displaced the centre of gravity. Although directional stability was deemed as 'satisfactory', it could be improved and Baur suggested that longitudinal stability be tested further with varying centres of gravity and stabilizer settings. The re-worked fitting of the nose landing wheel was found to make a definite improvement. Baur rated general performance to be fair, and recommended that several improvements be made, especially to certain mechanical defects in the brake system, radio system, pressurised cabin, manifold pressure and idling speed and in respect to control surface forces.[10] Also, some minor defects were found in the electrical cables of the intercom system.

Later flights in February by Baur and Gerhard Caroli revealed that the forces against the control surfaces were still too high, especially at high speed. Small defects were still present in the radio system and landing gear. The faulty trimming and controls revealed that an eventual change in the entire control system would inevitably be needed. Flights with two or three engines were found to be satisfactory, but in flights with the automatic controls it was found that the servos were not powerful enough to control such a heavy aircraft.[11]

On 12 February, Baur noted that reduced rudder deflection was sufficient for take-off, but the aileron linkage rods showed too much slack due to a failing of one of the servo controls. On the 26th, further tests were made on the still unreliable landing gear, three-axle control and flight on one engine. By using the rudder carefully, a speed of 600 km/h was attained, but Baur noticed that the airspeed indicator was showing a much higher speed than was actually the case, suggesting, in fact, a failure in the rudder mechanism. At 1,800 metres altitude, the required trim deflection of the rudder was established by switching off power in all but one engine.[12]

The flights made between 1-12 March were intended to assess direction-finding, stability, and to establish the front and rear centres of gravity during take-off and landing. It was discovered that in horizontal flight at cruising speed with the undercarriage extended, speed reached 290 km/h, but when the undercarriage was retracted, speed increased to 410 km/h.

'*Before any further flights are made,*' Baur wrote, '*I recommend that the following alterations be made:*

Install rudder counter-balance horns
Move back the axis of the tailplane to reduce pressure
Install a small tailwheel
Change the steering column
Move the EIV unit on the underside of the fuselage away from the running engines (as per the Me 323)
Fit new main wheels to allow greater weight.'[13]

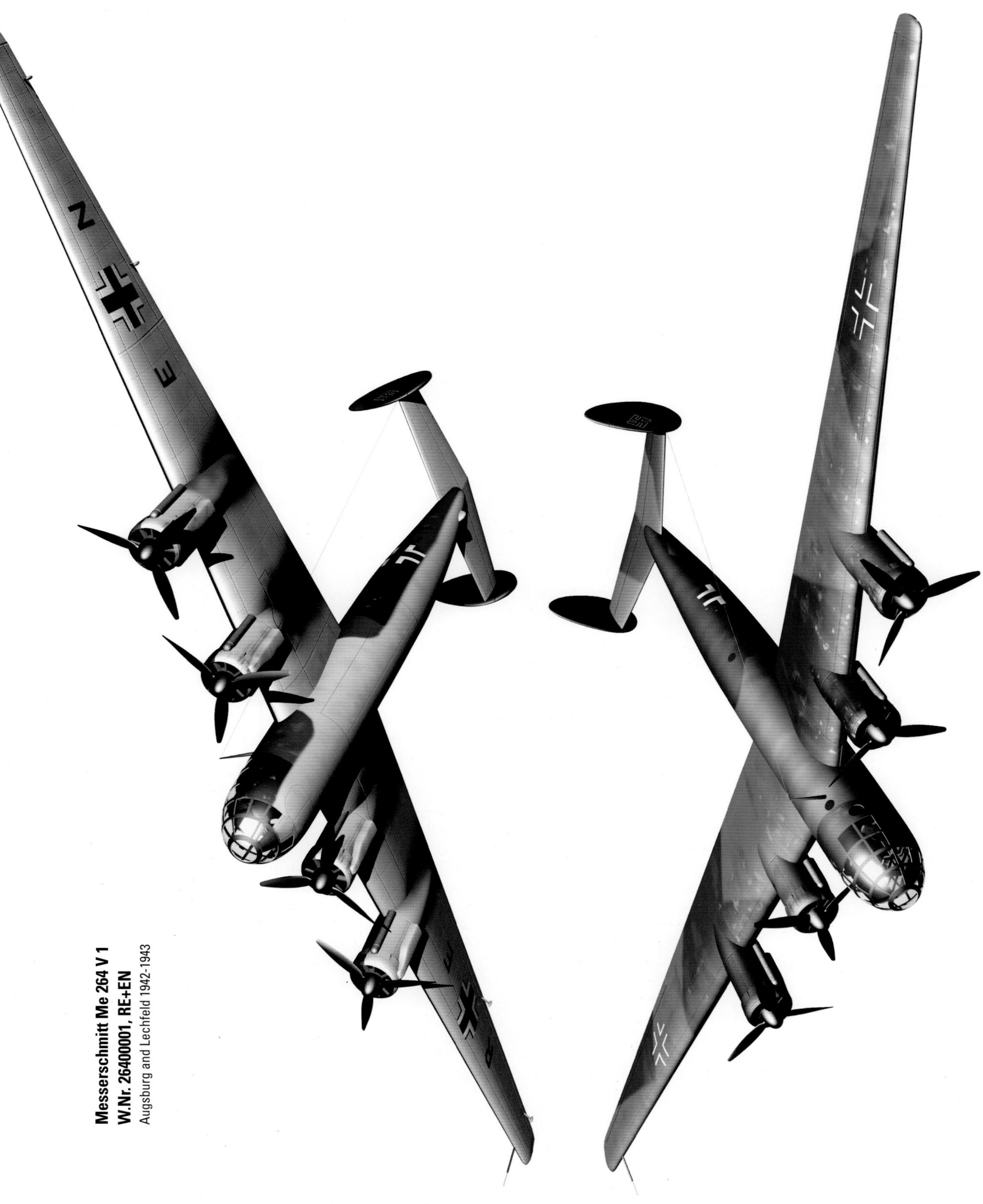

Messerschmitt Me 264 V 1
W.Nr. 26400001, RE+EN
Augsburg and Lechfeld 1942-1943

Above Detail view of the cockpit and starboard wing with its Jumo 211 engines. When retracted, the nosewheel was covered by three doors, two of which are visible here.

Below Messerschmitt personnel gather around the Me 264 V 1 at Augsburg or Lechfeld, with particular interest focused on the starboard in-board Jumo 211 engine. This photograph gives a good impression of the low profile of the aircraft; note that the tall Luftwaffe officer wearing a greatcoat near the cockpit stands over half the height of the aircraft. The civilian in the hat just visible standing adjacent to the Luftwaffe officer is probably test-pilot, Karl Baur.

Above and below Technicians inspect the engines of the Me 264 V1. Note the heavily oil-streaked cowlings (above). The figure wearing the hat in the cockpit in the photograph above is probably Karl Baur, and the man half in shadow in the centre of the group of three men standing directly below the cockpit appears to be Gerhard Caroli.

Above and below A small group of curious civilian personnel inspect the Me 264 V 1 at close quarters on a clear day at Lechfeld in 1943.

On 4 March 1943 a check of the auto-pilot system was scheduled; however, after 15 minutes of flight, the right inboard engine began to stream smoke and power had to be cut. Fortunately, there was never any danger to the test crew or the aircraft.[14] A replacement engine was fitted.[15] By this stage, the Me 264 had undertaken 11 test flights totalling 12 hours' flying time.

Flight-testing was interrupted briefly between 5 and 8 March while a new starboard side, in-board engine was fitted. On subsequent flights the aircraft was tested for various aspects of stability and pilot control.

On 9 March, three flights – the 12th, 13th and 14th – were made lasting 70, 27 and 33 minutes respectively; however during one of these, the automatic flight control system suddenly failed.[16] On the 12th flight, the propellers were tested during climb and high-speed flight, but the starboard, in-board engine became overheated and had to be stopped in flight. On the next flight, *Flugbaumeister* Friebel accompanied Baur on an inspection and assessment flight for the RLM. However, little value was gained from this flight due to its short duration. On the 14th flight, it was not possible to feather the port, outboard engine. The starboard in-board engine had to be stopped due to increased oil temperature and there was also an oil leak. However, after 15 minutes, the propellers started to turn freely. Finally that day, the aircraft was tested to assess operational performance at low-level. Upon landing, it was found that the exhaust covers were damaged and these were not replaced for later flights.[17]

On 10-11 March, maintenance was carried out on a problematic engine and new starter motors were fitted and when this was completed, Baur took the Me 264 up on a flight lasting just over two and a half hours to test longitudinal stability, followed by a second flight in which the engines were feathered.

Between 13-22 March, the aircraft remained on the ground while a new control column was installed and replacement trim tabs for the rudders were fitted. By 15 March, the aircraft had flown 14 hrs 10 mins and made seven flights. Engine running time had amounted to 22 hrs 39 minutes for the left outboard unit; 22 hrs 21 mins for the left in-board; 20 hrs 15 mins for the right in-board unit (although this engine had been replaced on 9 March and the new unit had been run for 4 hrs 52 mins); and 22 hrs 31 minutes for the right outboard unit.

Dipl.-Ing. Witte also conducted a thorough test of the hydraulics system.

In general terms, the more recent test flights using only two or three engines compared unfavourably with similar earlier tests.[18]

In this initial phase of the Me 264 flight test programme, the RLM instructed the Dornier firm to become involved with the project in order to assist Messerschmitt's undercapacity. However, during a conversation with Milch on 12 February 1943, *Oberst* Georg von Pasewaldt, a former competition pilot from the 1930s and now the head of the *Amtsgruppe Entwicklung* (GL/C-E – the development section) within the *Technisches Amt*, complained: 'We discussed the long-range question regarding the Me 264 and gave the machine to Dornier. The Dornier company's staff were very mistrustful and in any case have such a workload that they can't handle it. We then took it back, because a reluctant approach is not useful. Besides the Me 264 is so far nothing more than a flying mock-up. They still can't make anything of it. What we are supposed to do next with a project like that is causing us serious problems…'[19]

During a conference ten days later, Pasewaldt somewhat unwisely indicated to *Reichsmarschall* Göring that the first series production of the Me 264 could materialise as early as mid-1944 – this, despite the fact that it was anticipated that bringing the Me 264 into production would consume no fewer than 75,000 hours of construction time and that the first of the 30 pre-production aircraft was supposed to have been operational by the autumn of 1943. However, neither Blohm & Voss, Siebel, Focke-Wulf, Messerschmitt or Weser Flugzeugwerke had sufficient production capability for such a commitment.[20]

In what seems a complete contradiction to his statement to Milch on 12 February, Pasewaldt appears to have towed a different line when he attended an RLM conference in Berlin on transatlantic and long-range aircraft less than a month later, on 5 March. The conference was also attended by *Oberst* Karl-Henning von Barsewisch, the *General der Aufklärungsflieger* and *Oberstleutnant* Rowehl, the commander of the *Versuschsstelle für Höhenflug*. At one point the discussion turned to the massive six-engined Blohm & Voss BV 238 flying boat:

Von Barsewisch: *The Navy keeps bringing it up. They say it can refuel every so often from a U-boat tanker and stay out there for weeks. [Laughter rippled around the conference room]. No, seriously, they emphasize that! For that reason it's not being shelved. But now they're coming round to the idea that only a fast land-based aircraft is going to be of use for shipping reconnaissance. Therefore we just have to find a way to get the Me 264 into service.*

Pasewaldt: *The Me 264 must be built, we want it at all costs. But the thing is taking a long time. Meanwhile we have to make full use of the Ju 290.*

Oberst Karl-Henning von Barsewisch (left), the General der Aufklärungsflieger – Commanding 'General' of the Reconnaissance Arm – seen here with Hauptmann Erwin Fischer (centre) and Hauptmann Bergen of Fernaufklärungsgruppe 5. Von Barsewisch had served in the First World War as a Gas Protection Officer and a cavalry officer. He joined the Luftwaffe Reserve in October 1935. He then held various staff positions before being appointed commander of Aufklärungsgruppe 123, and then successively as Operations Officer with the Luftwaffe liaison detachment attached to the 4. and 18. Armee and 2. Panzer Armee. He was appointed General der Aufklärungsflieger on 28 November 1942, a position he held until the end of the war. He was promoted to Generalmajor on 1 November 1943 and awarded the Deutsches Kreuz in Gold on 10 January 1944. In early 1943, von Barsewisch supported development of the Me 264 as a maritime reconnaissance aircraft, until 1945, when newer technology offered more opportunities.

Petersen: *I would like to warn against exaggerated optimism over the Me 264. It needs 1,600 metres of runway with rocket assistance and 2,400 metres without. That is crazy!*

Rowehl: *But it can fly more than 10,000 km…*

Von Barsewisch: *Anyway, it seems from what was said in the talk with Dönitz that the U-boat war stands or falls by reconnaissance aircraft. This has been explained to the* Führer, *he has approved it and the* Reichsmarschall *has promised to fulfil the request. Undoubtedly, we must have a reconnaissance aircraft which can match the demands for mid-Atlantic cover.*[21]

In his view, *Admiral* Dönitz, and the *Seekriegsleitung*, regarded the Me 264, and later, the Ta 400 from Focke-Wulf, as more preferable aircraft with which to conduct long-range maritime reconnaissance missions than the existing four-engined Blohm & Voss BV 222 flying boat. His opinion had probably been influenced by the failure of refuelling trials with the BV 222 involving U-boat tankers. Dönitz' problem was that the Ta 400 was not expected to appear before 1946, and thus for the foreseeable future, the *Kriegsmarine* would have to rely on the Ju 290, Ju 390 and He 177 to provide interim maritime reconnaissance capability.[22] Yet it would not be until 3 May that the SKL noted that the *Luftwaffe* had finally decided to '… plan for a long-range reconnaissance aircraft suitable for Atlantic missions.' Milch and Jeschonnek were agreed that the Me 264 was unlikely to make its operational debut before 1944 or even 1945, and a number of technical experts were instructed to explore urgently the possibility of fitting auxiliary tanks to the Me 264 to increase range still further.

Five days after the Berlin conference, on 10 March 1943, the Allies were offered their first glimpse of the Messerschmitt bomber when a Mosquito PR IV reconnaissance aircraft, DZ.364, of the RAF's 540 Squadron crewed by F/O A. Stuart and F/Sgt M. Pike took off at 1330 hrs from Benson in Oxfordshire on a photo-reconnaissance mission to the Munich-Augsburg area. The night before, RAF Bomber Command had despatched 264 heavy bombers to bomb Munich, and Stuart and Pike were sent out to make a post-raid reconnaissance. As the Operations Record Book of 540 Squadron records, the Mosquito made course over the English Channel, east, towards the Belgian coast:

> '*Ostend 1500. South east of Munich. Long-range tanks empty after 55 minutes flying. Visibility very bad. Very hazy. Winds stronger. 1550 Strasbourg aerodrome. Photos. 1630 Augsburg. Photos. Then Munich two runs and at least one fire still burning. Other targets were aerodromes Lechenfeld* [sic – author], *Le Treport. Photos. Landed base 1840. Unfortunately petrol film over rear camera – otherwise OK.*'[23]

As a result of their pass over 'Lechenfeld', Stuart and Pike returned with the first photograph of the Me 264 which, according to the RAF's Photographic Interpretation Unit at Medmenham, showed '*… a very large aircraft… very indistinctly visible at Lechfeld on 10 March 1943…*'[24]

The Me 264 was no longer a secret – but neither was it close to becoming an operational aircraft. Three days after Stuart and Pike flew over Lechfeld, Göring vented his frustrations at an assembly of aircraft manufacturers and senior RLM officials attending an aircraft production conference at his opulent country estate at Carinhall: 'Gentlemen, I have called you together again today to speak about the entire situation on the technical side of the *Luftwaffe* and to inform you of my views on the subject, and most importantly, those of the *Führer*. It would have been very agreeable if I could have commenced my remarks today by acknowledging and thanking you for your efforts. However, I find myself unable to do this if I am to continue speaking frankly – quite the reverse. I can only express to you my absolute bitterness about the complete failure which has resulted in practically all fields of aeronautical engineering – bitterness too that I have been deceived in the past to an extent such as I had experienced only in variety shows at the hand of magicians and illusionists – such has been the hocus-pocus which everybody has used to take me in. Whenever future problems were under discussion, everyone already

Below and below right The wind tunnel model of the Focke-Wulf Ta 400 six-engined bomber and reconnaissance project. With a 42 m wingspan, this was the favoured contender to the Me 264, but in the spring of 1943, it was not envisaged to become available until 1946. Power was to be provided by six 1,700 hp BMW 801D engines, and two pressurized sections – in the forward and rear sections – were to accommodate the pilot and co-pilot and the operator of the rear rotating turrets. Two observation domes were mounted on the fuselage sides for the remote-controlled gun positions. The tail was of a twin fin and rudder design, and the tailplane incidence could be adjusted hydraulically. 27,000 litres of fuel was to be carried in no fewer than 32 tanks, 12 in each wing and 8 in the fuselage. A single nosewheel retracted to the rear and four main wheels retracted forwards. Basic armament was planned as one remote-controlled turret beneath the fuselage (two 20 mm MG 151 cannon), two remote-controlled turrets on the upper fuselage (each with 20 mm MG 151 cannon) and a remote-controlled tail turret mounting four MG 151 20 mm cannon. Range was projected at 4,800 km, with a bomb load of 10,000 kg at a maximum speed of 535 km/h.

had the most fantastic things ready and it was then only a matter of production before they could be brought into service...

'I well remember that at Augsburg – it was exactly a year ago – I was shown an '*Amerika Bomber*' that really called for nothing more than to be put into mass production. It was to fly to the east coast of America and back, from the Azores to the American west coast and also carry a lot of bombs. I was told so in all seriousness. But in those days I was still so trusting, I half believed that something like this was possible.... Fun has been made of the enemy's backwardness and his slow four-engine crates... Gentlemen, I would be extremely happy if you could produce one of these crates in the immediate future. I would then at least have *one* aircraft with which something could be achieved...

Göring piled on the pressure: 'I cannot do much with what you are giving me at present. Even if the He 177 is produced, what am I to do with it? It can hardly get its nose past the hangar doors and cannot even reach Glasgow. The same applies to the Ju 188. Even our fighters can reach London... It is enough to drive one to despair! Year after year has gone by and you have plodded away at the same old things. First an engine is drilled out a bit more... then the wing-tips are snipped off or something else is done. But a new aircraft, which can really do the job does not materialise. Tell me the name of such an aircraft!' [25]

Göring also expressed his displeasure at the inadequacies of the Ju 88: 'I recall the marvellous circles they drew on their charts for me showing the radii – how this aircraft could cruise up and down the west coast of Ireland attacking enemy shipping lanes. But we *still* have not got any such aircraft!'[26]

At the end of the month Göring vented his frustrations to Werner Baumbach, this time regarding the continuing delays to the Heinkel He 177; he commented wistfully: "I was promised a big bomber, the He 177, which should have been in service a year ago. But when it was tried out, it had catastrophic losses, and not in action either. A year has gone by, and when some sort of a thing comes along in a year or so it will probably prove to be hopelessly out of date."[27]

Matters were not helped when, on 23 March 1943, following its 17th flight – to test steering and general control – lasting just under two hours, the left oleo-leg of the Me 264 V 1 '*Sudeten*' broke during landing at Lechfeld. This was probably due to the undercarriage mechanism not engaging and locking properly. Karl Baur recalled: '*Landing took place with crosswind of approximately 6 m/s. At touchdown the left-side landing gear broke. I was able to keep the aircraft in a straightforward position for about 600 metres. The left wing had ground contact at that point and then the aircraft made a 180 deg turn to the left. The right landing gear collapsed as a result of it, and the front wheel was sheared off. The propellers were damaged on all four engines. The tail of the aircraft had folded and the vertical fin and left outboard wing had been ripped off.*'[28]

It was a blow to the already dubious reputation of the Me 264 and the crash delayed further testing considerably. However, while repairs were being carried out, work was finished on the new steering column and the aircraft was fitted with a reinforced wing skin, a modified nosewheel drive and new radio equipment. Further modifications at this time, which would last until 21 May, included the fitting of an improved emergency tail skid, a new tailplane and nosewheel and the installation of four new Jumo 211 J engines.[29]

On 21 April 1943, a meeting took place of the key design, development, engineering and testing personnel associated with the Me 264 at Messerschmitt Augsburg, to decide on the necessary changes to be made to the flight development programme, such as modifications to the hydraulics, landing flaps and rudders, undercarriage linkage and the apparently insolvable problem of pressure on the control surfaces.[30]

At a meeting of the RLM Development Committee on 27 April, it was decided to place faith in the tried and trusted Ju 290. This aircraft had by now at least seen action as a transport at Stalingrad with K.Gr.z.b.V 200. As for the Me 264, still languishing at Lechfeld undergoing repairs following its landing gear collapse the previous month, *Generalfeldmarschall* Milch remained far more dubious. He turned to *Oberst* Alpers of GL/C-B2, the *Technisches Amt* Procurement section, and enquired as to progress with the production programme:

Alpers: Thirty machines were ordered... Finding room for an order of this size is causing difficulties in Germany because we are already beyond capacity. Junkers suggested we could take over a supplier and use the works for assembly as well. I would like to try it. Junkers suggested the works at Toulouse.

Milch: Doesn't sound too good to me. Isn't there anything else?

Alpers: The problem is the size of the aircraft...

Milch: I think there are still many obstacles and difficulties to be overcome before the four-engined Me 264 is a practical proposition. Here again we have the same construction difficulty [as the Ta 400]. What worries me is that it can only get off the ground using rockets. I am no friend of such methods. Of course, I don't blame people for using them to break world records, but they shouldn't be for everyday use. Everything depends on good luck. If just one of the eight rocket boosters fails, the take-off will come to grief and probably leave us with a write-off. This is the spanner in the works. Things are even more difficult with the six-engined version. We would be over the moon if we could use it to bomb America. With a bombload, America is just about within its radius of action of 14,000-15,000 km. But it is borderline, and we have to weigh up whether it isn't better to have a seaplane do it in relays.

Above On 23 March 1943, following its 17th flight, the port landing gear leg of the Me 264 V 1 broke during landing at Lechfeld. It was a blow to the already dubious reputation of the Me 264 and the crash delayed further testing considerably.

Below Mechanics prepare to open the access panels to the nacelle of the outer port engine of the Me 264 V 1 under the watchful eye of other Messerschmitt personnel. The man to the far left wearing a hat and suit is probably Karl Baur.

Above With the Me 264 V 1 parked on the grass at Augsburg, a mechanic walks away as all four engines run up.

Left A rare view taken from the rear of the Me 264 V 1 at Lechfeld in 1943, with two mechanics working on the outer port engine.

Below A map of the world, dated 12 May 1943 indicating the difference in range capabilities between the 'Fw 400 3 ton bomber' and various Me 264 variants. Note that according to these Messerschmitt calculations the Focke-Wulf bomber would fall far short of the United States East Coast compared to the Me 264.

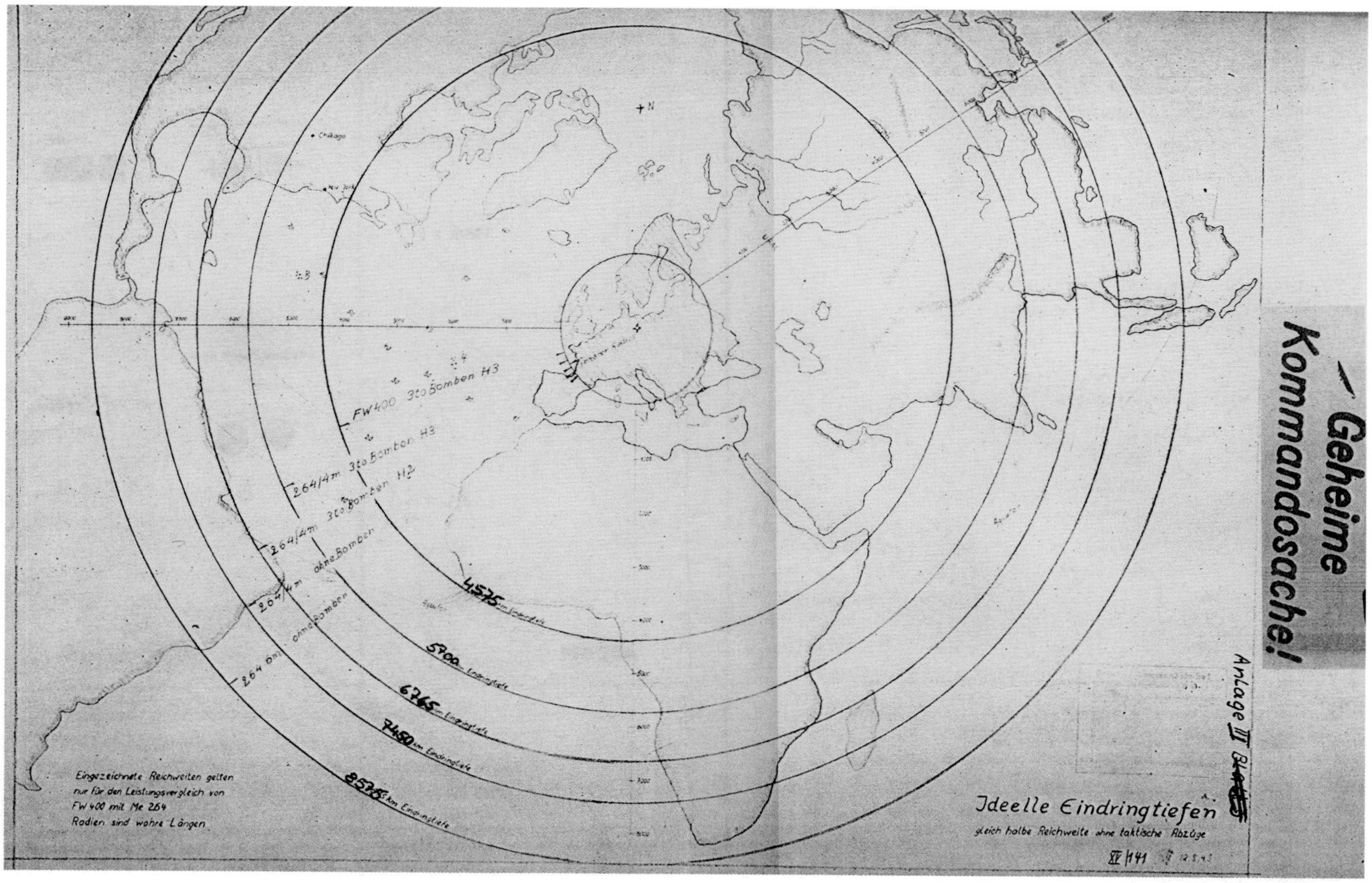

Oberst **Pasewaldt:** We must not pass over the Ju 290. She must fly in the form in which she has been reactivated and for such time until a successor or improved version comes on the scene... In the nature of things the Ju 290 will continue in the shape of the Ju 390. Consequently, we must not pass over the Ju 390...

***Generalleutnant Dipl.-Ing.* Wolfgang Vorwald** (Chief of the *Technisches Amt*)**:** I can only second the verdicts of von Pasewaldt and Petersen. The Ju 290 and 390 are ready. How many Ju 390s can be built we can leave until later... As for the Me 264, they should complete the prototypes and obtain experience with them...[31]

Further 'experience' would not be possible until repairs were completed at Lechfeld on 22 May. In the meantime, there was shock at Augsburg when an unexpected teletype message arrived from the RLM ordering that further work on the Me 264 should be abandoned. This was all the more astonishing because it had only been a week before that the company had received *Generalleutnant* Vorwald's instruction that work on the prototypes *must* be completed![32]

As a further example of paradox, by the spring of 1943, the Messerschmitt *Projektbüro* at last completed its basic aerodynamic calculations for the Me 264 with a lengthened nose and a new rear turret. There was also still much calculation and research work going on behind the scenes on the feasibility of a six-engined version of the '*Sudeten*'. Despite the fact that in late April Messerschmitt confessed that the design for a six-engined Me 264 represented the 'limit of development possibilities' and that, as such, flying qualities of the machine did not commend it for operations, in mid-May the company's recently appointed Chairman, Friedrich-Wilhelm Seiler, together with *Dipl.-Ing.* Seifert and *Dipl.-Ing.* Hans Hornung of the *Vorprojektbüro* (the Preliminary Project Department) busied themselves preparing detailed performance and technical comparisons between a normal configuration Me 264 with four BMW 801 E engines, the similarly equipped six-engined bomber proposal from Focke-Wulf – the Ta 400, and a proposed six-engined Me 264, also fitted with BMW 801 Es.[33] (See Appendix 1)

They concluded that while the wingspan of the Me 264 would and could be increased together with the control surfaces, the basic design of the Me 264 should not be changed unnecessarily. Furthermore, unlike the Focke-Wulf project, the Me 264 existed and would offer comparable – if not better – performance for less development time. The Focke-Wulf would require more materials such as strategic

Below A Messerschmitt design office drawing showing the interior layout of the crew accommodation of the Me 264. Note that the design for the nosewheel is still based on the original proposal in that it retracted horizontally under the pilot's position. The drawing shows the aircraft fitted with BMW 801 engines.

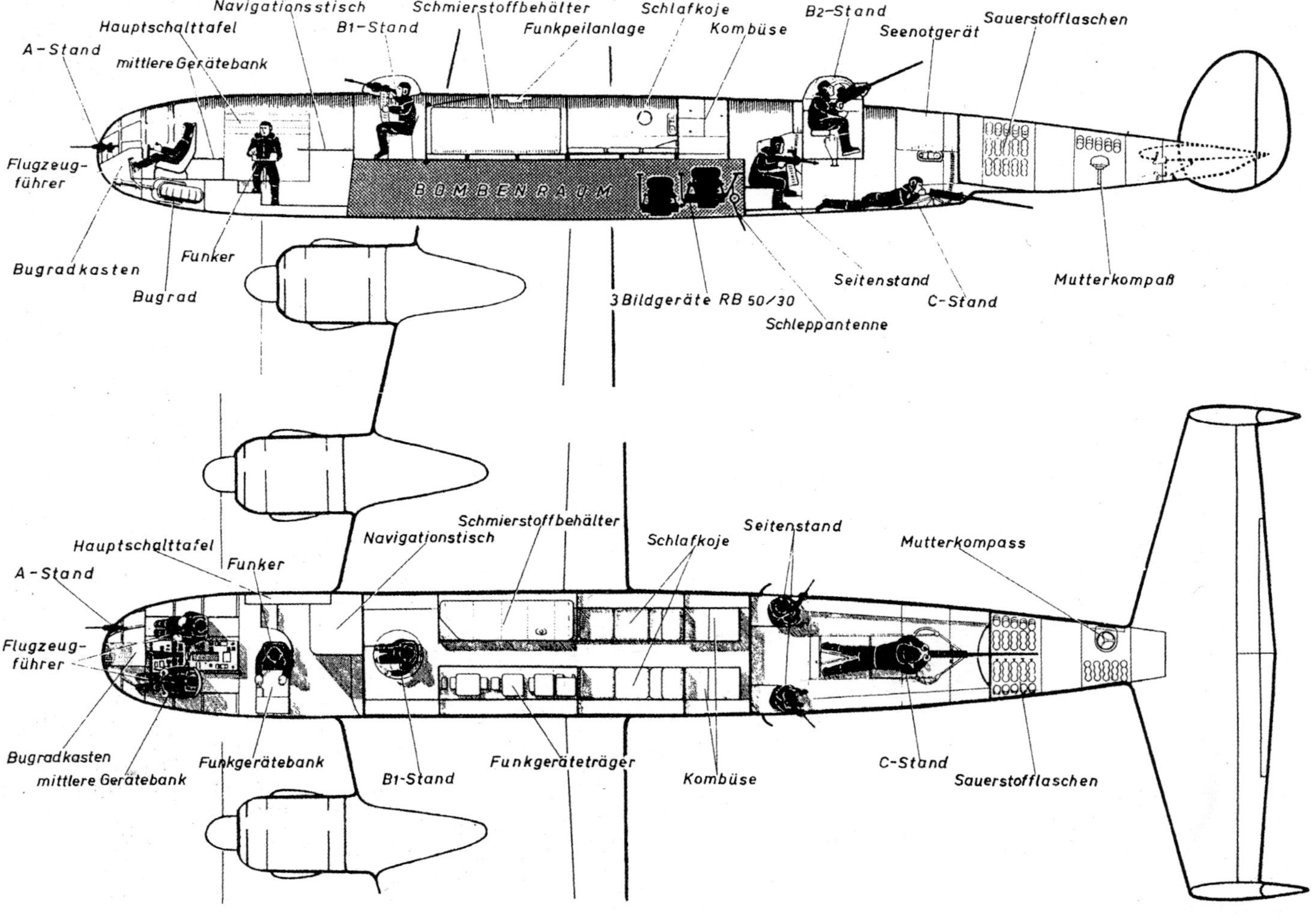

metals and rubber for its fuel tanks and other areas. The Messerschmitt team also noted that the Ta 400 would consume 60 per cent more fuel than a four-engined Me 264 – or eight tons more fuel per aircraft for a given long-range mission – and even more fuel than a six-engined Messerschmitt.

Left On 22 May 1943, the General der Jagdflieger – Commanding General of the Fighter Arm – Generalmajor Adolf Galland, visited Lechfeld to inspect and fly an Me 262 jet fighter prototype. As he concluded his flight, he passed the Me 264 V 1 as it came in to land and used the bomber to make his 'first exercise in jet fighter attack'. Galland (left) is seen here on the day talking to Karl Baur (centre) and Fritz Wendel, both of whom flew the Me 264. Wendel flew the Me 264 V 1 on 2 June 1943 and noted 'serious problems'. He had served as a Luftwaffe instructor before joining Messerschmitt in 1936 as a production test pilot. On 26 April 1939, he piloted the Me 209 V 1 which set a new World Absolute Speed Record at 755 km/h and was later heavily involved in the Me 262 test programme.

Meanwhile testing had finally resumed at Lechfeld and on 22 May 1943, the airfield received a visit from the *General der Jagdflieger, Generalmajor* Adolf Galland, who had arrived to inspect and fly one of the new Me 262 jet fighter prototypes also undergoing trials there. As Galland swept through the sky above Lechfeld in the Me 262 V 4, causing him later to make his famous 'pushed by angels' remark, he suddenly came upon a large aircraft making its landing approach; as he recalled in his memoirs: '... *the four-engined Messerschmitt flew over Lechfeld, and so became the object of my first exercise in jet fighter attack.*'[34]

More testing took place between 25 May and 5 June, the latter day seeing the Me 264 fly six times for a period of 12 hours and 16 minutes. These flights were made mainly to assess the functioning of the newly configured ailerons as well as gun turret mock-ups which had been installed. However, once again Baur encountered the old problem of excessive forces on both the ailerons and rudders. Furthermore, because of its high wing-loading factor, the Me 264 was not easy to handle and required the highest degree of skill during take-off, something that even a pilot of Baur's experience and capability found a challenge.[35]

Another persistent problem was the tendency for the nosewheel to jam on retraction, such as when Messerschmitt's other accomplished test pilot, Fritz Wendel, took the controls of the Me 264 on 2 June. Wendel noted 'serious problems' in this area. On the 10th, *Flugbaumeister* Böttcher of the RLM Aircraft Development Department also flew the machine and, due to the warmth of the summer weather, complained of intolerably excessive heat in the extensively glazed 'greenhouse'-style cockpit.

Still the shadow of the Me 264 fell across conference tables at the very highest quarters. With its potential range, Hitler and *Admiral* Dönitz saw it as a vital aid in the burgeoning U-boat battle against the Atlantic convoys. 'There can be no talk of a let-up in submarine warfare,' Hitler said at one war conference in mid-1943. 'The Atlantic is my first line of defence in the West, and even if I have to fight a defensive battle there, that is preferable to waiting to defend myself on the coast of Europe. The enemy forces tied up by our submarine warfare are tremendous, even though the actual losses inflicted by us are no longer great. I cannot afford to release these forces by discontinuing submarine warfare...'[36]

In June however, *Admiral* Dönitz, whose boats were making near-suicidal attempts to attack the convoys, told Hitler: 'The declining figures of sinkings in the U-boat war can only be made good by making more use of the *Luftwaffe*.'[37]

The *Führer* craved more success at sea and put pressure on his air force to deliver the necessary support to the U-boats. On 15 June 1943, *Oberst* Viktor von Lossberg, the General Staff Officer to the Chief of the RLM *Technisches Amt*, presented another report dealing with possible raids against US targets. He saw the Blohm & Voss BV 222 undertaking such missions, supported by U-boat tankers in the mid-Atlantic which

Above In the summer of 1943, Oberst Viktor von Lossberg, the General Staff Officer to the Chief of the RLM Technisches Amt advocated using Blohm & Voss BV 222 flying boats to conduct raids against American targets, using U-boat tankers to refuel them in mid-Atlantic. Simultaneously to this, von Lossberg also became involved in developing the 'Zahme Sau' freelance night fighting system, an effective tactic for night fighters which was to become standard practice for the Nachtjagd for the duration of the war. Von Lossberg is seen here with the rank of Major.

Left The enormous Blohm & Voss BV 222 Wiking flying boat was Oberst Viktor von Lossberg's choice as the aircraft best suited to make raids aganst the United States on the basis that it would be refuelled in the Atlantic by U-boat tankers. This machine, the BV 222 V 7, X4+CH, was the fourth of its type fitted with Jumo diesel engines and it entered service with Aufklärungsstaffel (See) 222 in August 1943.

Left Lechfeld, summer 1943: the awesome span of the Me 264 V 1 photographed from the outer section of the port wing, looking across the wings. Note the mock-up gun turret and the observation window in the fuselage centre section.

Below A frontal view of the Me 264 V 1 taken at Lechfeld during the early summer of 1943. Two turret mock-ups have been installed on top of the fuselage, with a view to installing remotely-controlled turrets fitted with 13 mm MG 131 Zwilling combinations.

Left and below Two views of the Me 264 V 1 fitted with mock-up gun turrets designed to replicate remotely-controlled 13 mm MG 131 Zwilling combinations.

would offer refuelling. Surprisingly the SKL supported this risk undertaking, but the *Luftwaffe* General Staff showed little interest. That day one RLM engineer recorded: '*What we have on hand at the moment for long-range reconnaissance are the Ju 290 and the He 177 which can do the job for a while. They will be replaced by the Ta 400. Accordingly this leaves the Me 264 in a poor position. I flew it recently and I believe it is the right thing to drop it.*'

Milch supported this view, advocating the Ju 290 and He 177 for Atlantic reconnaissance, to be eventually replaced by the Ta 400 with the Me 264 continuing as a development project only. The Me 264 production plans would therefore be cancelled.

On 8 July 1943, with German armoured forces heavily engaged in a massive pivotal battle at Kursk on the Eastern Front, Hitler took time away from the worries of the land war to meet once again with Dönitz at the *Wolfsschanze*. The Admiral presented the latest bleak situation report from his U-boats. Hitler again spoke of the four or six-engined Me 264, an aircraft whose range of up to 17,000 km he envisaged

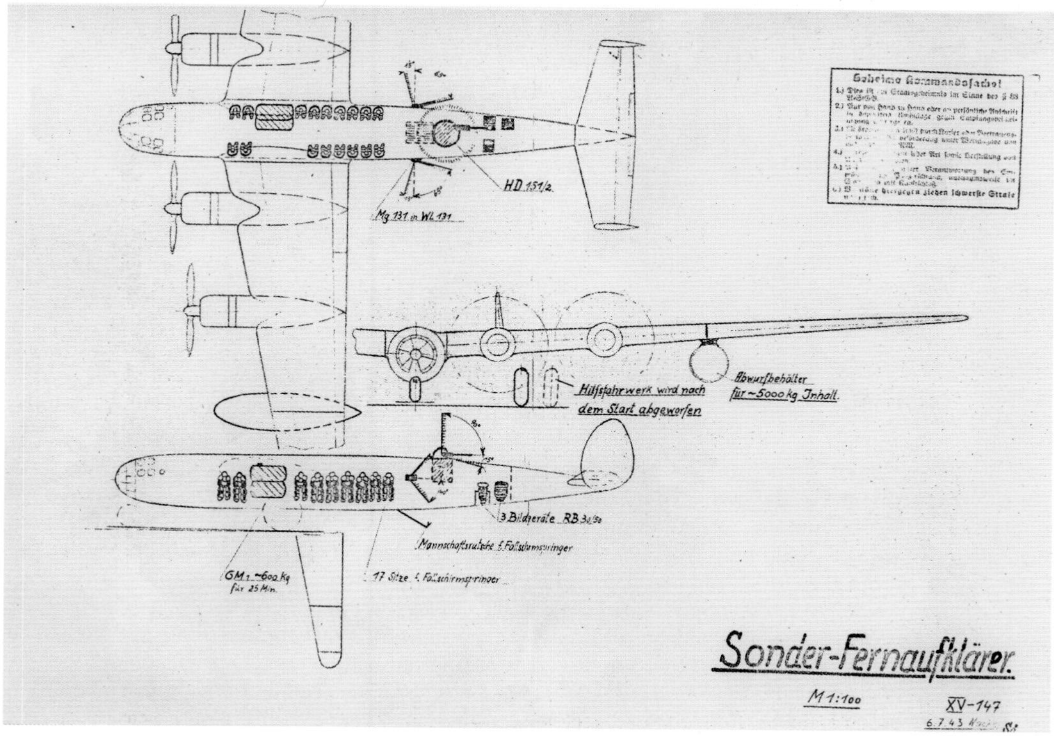

This page and overleaf A set of three Messerschmitt drawings dated 6 July 1943 showing proposed operational variants of the long-spanned Me 264.

Left The Sonder-Fernaufklärer – Special Reconnaissance – aircraft in which it was planned to carry 17 paratroopers or assault troops, as well as three RB 30/50 cameras. The aircraft was to be equipped with GM-1 nitrous oxide power boost and carry an additional 5,000 kg drop tank under each wing. Armament was to consist of a HD 151 upper fuselage gun turret and an MG 131 machine gun in waist positions either side of the fuselage. The aircraft would take off using jettisonable outer mainwheels.

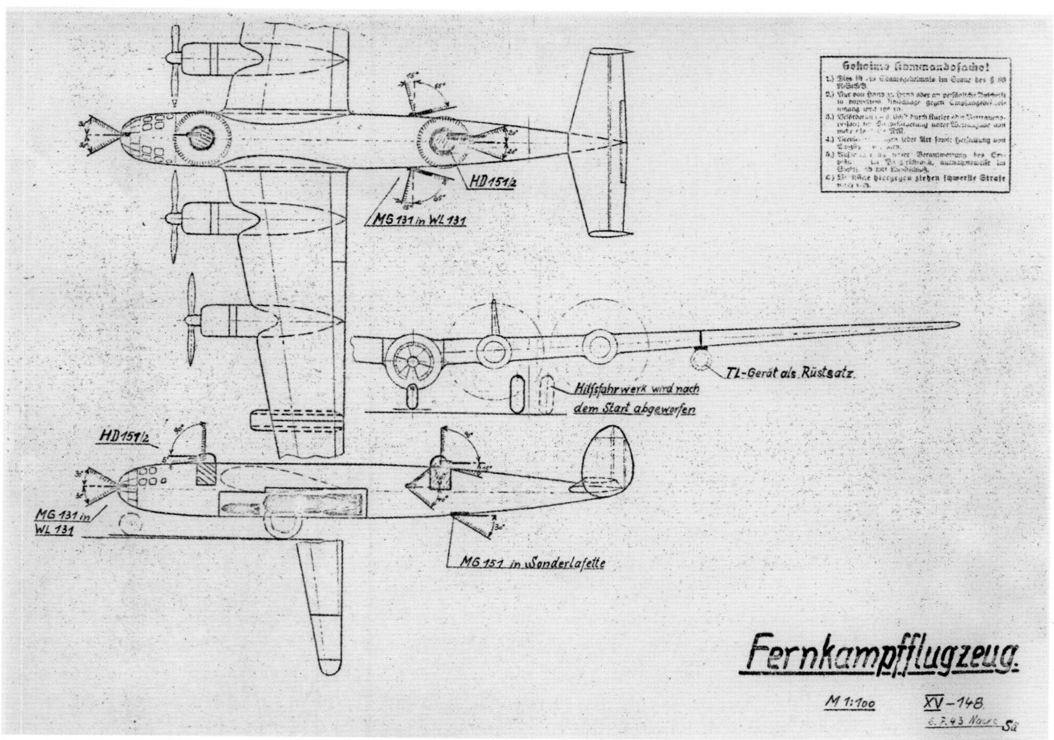

Left The Fernkampfflugzeug – or long-range bomber – was to have carried a heavy bomb load, possibly at least one SC 2500 in the bomb bay with other smaller ordnance. Note also the optional jet engine 'Rüstsatz' fitted to the underside of the outer wing. Armament was to comprise one MG 131 in the nose, two HD 151 Z turrets in the upper forward and rear fuselage sections, single MG 131s in waist positions either side of the fuselage and a single MG 151 remote-controlled barbette on the underside of the rear fuselage. The aircraft would take off using jettisonable outer mainwheels.

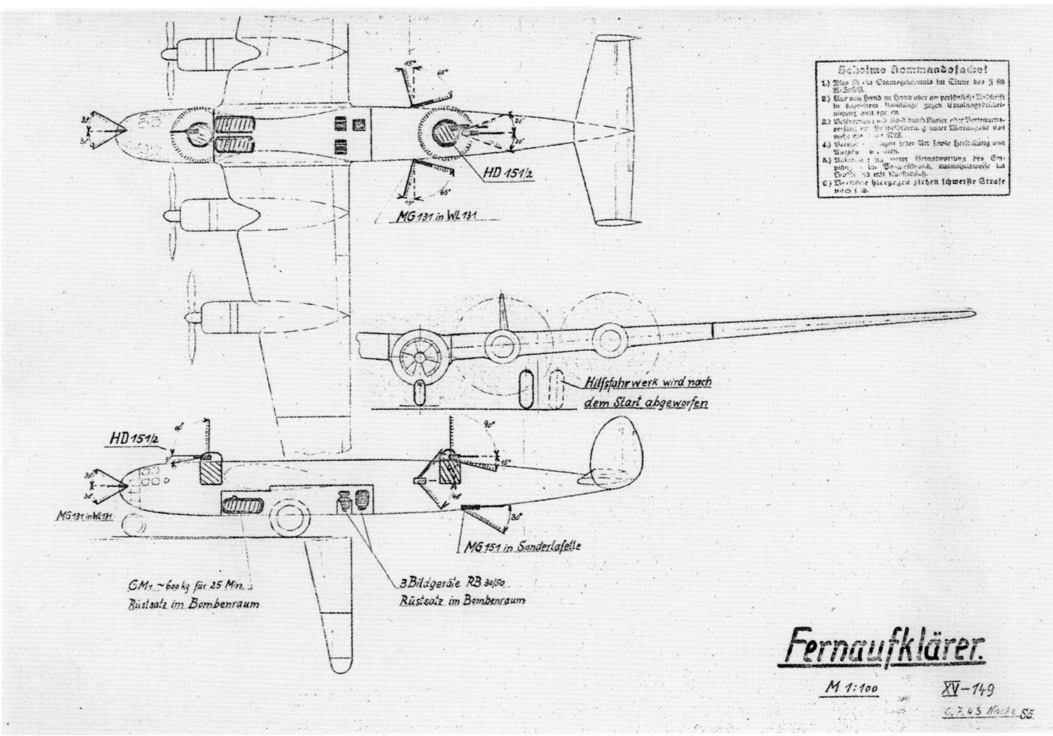

Right The Fernaufklärer – intended for standard reconnaissance missions – was fitted with three RB 30/50 cameras with defensive armament the same as the Fernkampfflugzeug. The aircraft was to be equipped with 600 kg of GM-1 nitrous oxide power boost for a duration of 25 minutes fitted in the bomb bay. This would allow for increased speed in an emergency situation. The aircraft would take off using jettisonable outer mainwheels.

operating not only as a maritime reconnaissance machine but also as a fast bomber: 'These are the machines,' he told Dönitz, 'which shall later cooperate with the submarines.'

The *Führer* assured Dönitz of his support for the continued production of the Me 264, but only for maritime deployment. He also conveyed this to the Messerschmitt company. At the same time he dropped his decision to mount bombing attacks on the East Coast of the United States because '... the few aircraft that could get through would only provoke the populace to resistance'; it seems Hitler had recognised the stubbornness of both the British and German civilian populations under aerial bombardment.

The following day, Milch reversed his earlier decision and agreed to continue the construction of the three Me 264 prototypes for the purpose of 'study' only, on the basis that the Weser aircraft factory be included in the production process. However, this latter plan never materialised.[38]

There is little doubt that Willy Messerschmitt took a direct hand in communicating with Hitler about the future development of his four-engined bomber at this time. There is also little doubt he painted a 'rosy' picture. The day after Hitler and Dönitz met, a conference took place at the RLM with *Oberst iG* von Pasewaldt and *Oberstleutnant* Ulrich Diesing of the RLM Technical Department amongst those attending. Pasewaldt reported: 'I would now like to mention the question of the Me 264, which will probably also interest *Oberst* Peltz. The matter has been put into a new light following the manufacturer's report to the *Führer*. The situation was that the *Herr Feldmarschall* Milch had taken the decision and reported to *Reichsmarschall* Göring that the Me 264 prototype was completed and that the Weser firm had capacity. It then appeared that the capacity at Weser was not sufficient because substantial additional requirements had arisen which could not be met. At the same time there arose the urgent need for the Fw 190 D, and now the Ta 152, which with regard to the general situation was very awkward for us. The remaining available capacity at Weser was given to Tank. Messerschmitt had made strenuous efforts to be allowed to look for a way to build the aircraft at his own plants. Since then, Messerschmitt has received no satisfactory reply to the question, which is to say that the question of where the aircraft is to be built is still up in the air. Since *Oberstleutnant* Diesing has now said that the Me 264 will be ordered after all, the question of where this aircraft is to be produced is naturally once again of extraordinary importance.'

Diesing: 'After *Professor* Messerschmitt visited the *Führer*, the *Reichsmarschall* asked me what the Me 264 was. I described the machine and added that, of the six planned prototypes, one was ready as a flying mock-up without armament and equipment and that parts had been taken out or were ready for fitting. Thereupon the *Reichsmarschall* pointed out that *Feldmarschall* Milch was of the opinion that this prototype must be properly flight-tested before a decision was taken about a big production run, basically because quite often with Messerschmitt's new aircraft a weakness appears requiring a big reconstruction job.'[39]

It is unclear from where Diesing had the impression that there were 'six planned prototypes.'

Meanwhile, throughout the summer of 1943, the designers and project engineers at the Messerschmitt *Projektbüro* continued their work into further development of the '*Sudeten*'. One such proposal was for a high-altitude bomber version with a take-off weight of 39,000 kg and a partly-jettisonable landing gear. It was to be equipped with four supercharged BMW 801 E/F radial engines, giving it a cruising speed of 640 km/h at 12,000 metres. As and when the Jumo 222 E/F engines eventually became available, the BMWs would have been replaced. Due to the required use of the aircraft, the design also included a

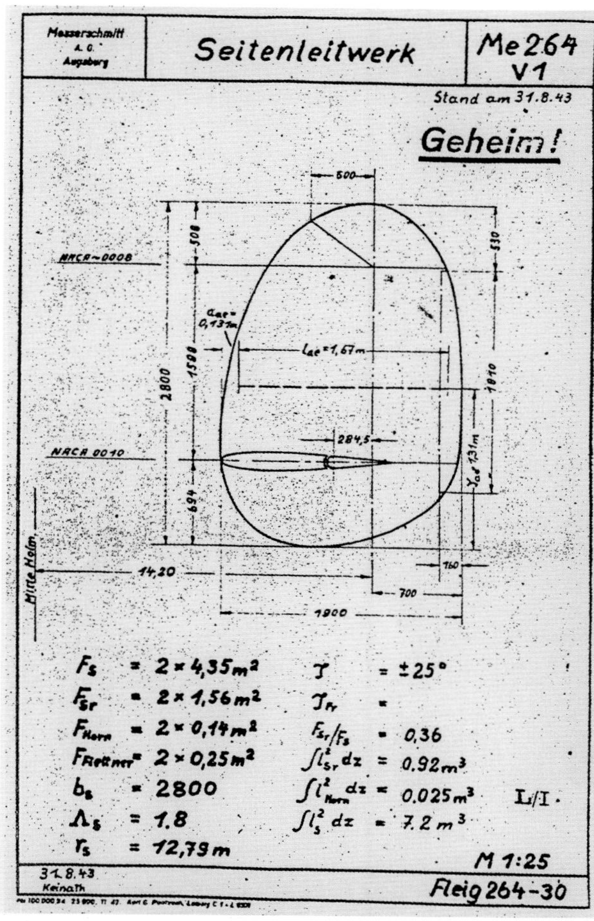

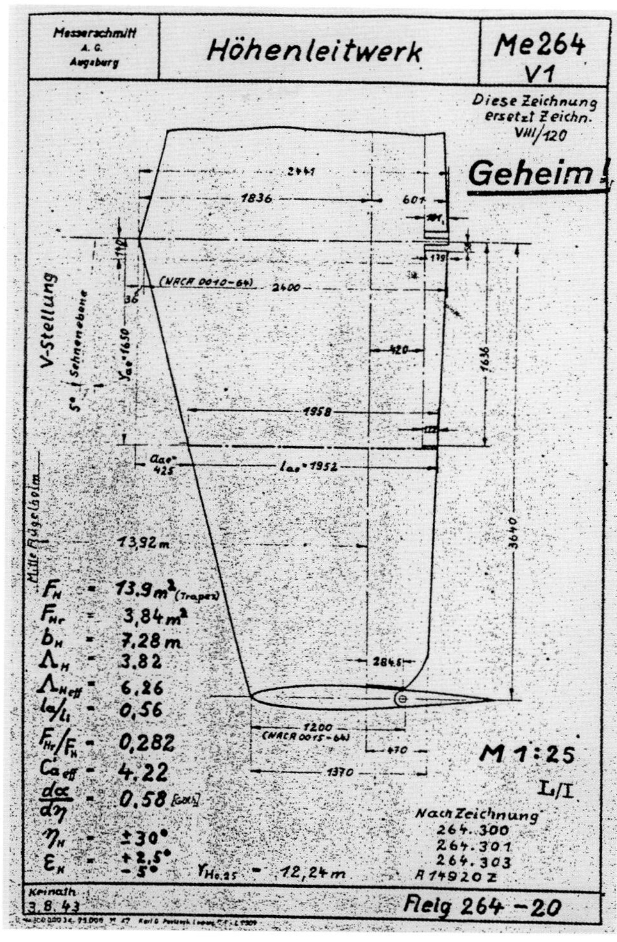

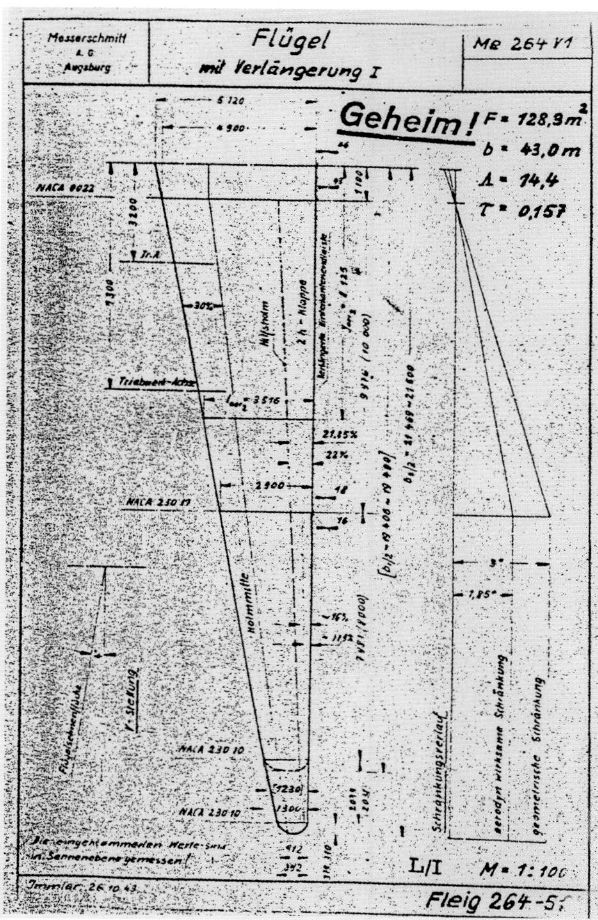

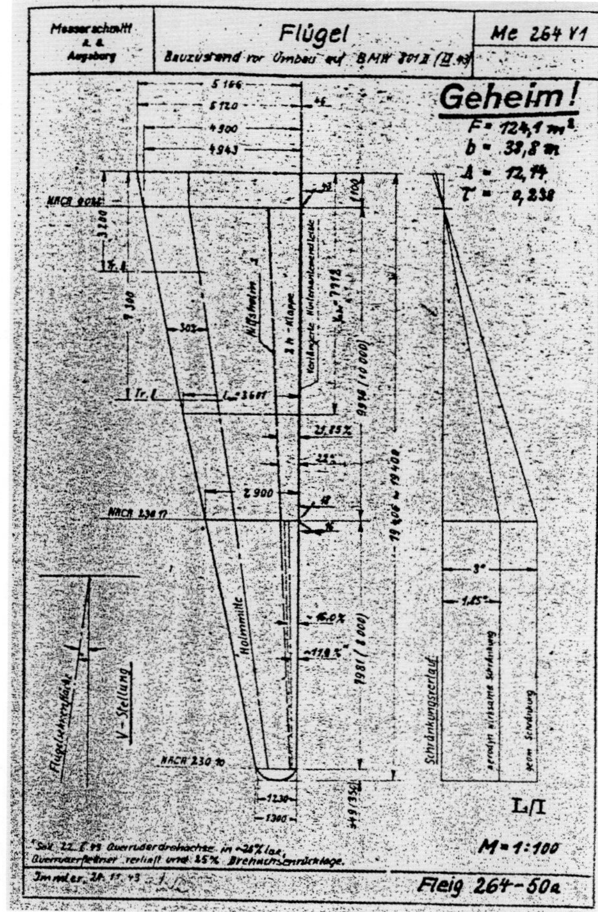

Four detailed dimensioned drawings of the tail assembly and the two proposed wing variations for the Me 264 V 1 produced between August and November 1943.

pressurised cockpit, although that was the only part of the machine which was to be so equipped. The rest of the aircraft remained unpressurised and so it would either have remained unarmed or its defensive armament was to have been remote-controlled.

For operations – presumably intended to be mounted against the United States – the bomber would have a minimum penetration depth of 3,500 km. However, to extend the range of the aircraft still further, the Messerschmitt designers came up with the idea of fitting two very large drop tanks to the undersides of the wings, each weighing 5,150 kg when full (see Chapter Five). Detailed plans were drawn up but by the time the factory was ready to build these in September 1944, work was abandoned.[40]

Work also progressed around the creation of 'optimal' piston engines with additional jet-powered propulsion units, or just jet or turboprop engines. A report prepared by BMW on 7 October 1943 proposed the BMW 801 GM, the BMW 803, and BMW 018 jet engine as well as the turboprop PTL 028 for high-altitude long-range operations. BMW concluded that the BMW 803 with its lower fuel consumption would be far superior to the BMW 801. With two BMW 018s a higher operational speed and a faster rate of climb could be attained, though there would be higher fuel consumption. Unfortunately, there was little real prospect of fitting the Me 264 with such powerplants because its airframe was already too heavy.

BMW calculated the following performance figures:[41]

Me 264 long-range bomber – maximum speed fitted with:

4 x BMW 801 TM	600 km/h at 10,515 m
2 x BMW 803	639 km/h at 12,500 m
2 x BMW 018	779 km/h at 7,010 m
2 x BMW 028	805 km/h at 7,010 m

Range:

4 x BMW 801 TM	(27,276 ltrs fuel) 6,219 km
2 x BMW 803	(27,276 ltrs fuel) 8,883 km
2 x BMW 018	(23,866 ltrs fuel) 5,423 km
2 x BMW 028	(23,866 ltrs fuel) 6,501 km

Me 264 long-range reconnaissance aircraft/bomber fitted with 4 x BMW 801 TG and 2 x BMW 018:

Maximum speed with:

4 x BMW 801 TG	568 km/h at 6,400 m
2 x BMW 018	721 km/h at 7,467 m
4 x BMW 801 TG and 2 x BMW 018	793 km/h at 7,315 m

Rate of climb at sea-level at 53,388 kg with 4 x BMW 801 TG and 2 x BMW 018 = 810 m per min.

On 29 October 1943, a detailed report was issued by Fidelis *Freiherr* von Stotzingen and *Herr* Huber of BMW Flugmotoren GmbH in Munich on the performance enhancement offered to the Me 264 by the combination of four BMW 801 TG engines supplemented by two BMW 018 turbojets for the purposes of climbing and high-speed.[42] Under such a configuration, no external tanks would be fitted and a fuel load of 20 tons would be carried.

There had originally been the suggestion that the Me 264 could be fitted with just two BMW 801 TG engines, but von Stotzingen and Huber had concluded that in such an instance with a full fuel load and even without external tanks, the aircraft would not be airworthy. However, the BMW engineers also believed that any disadvantage in weight caused by using six engines was compensated by the increased level of safety should one engine fail or be put out of action.

The envisaged weight distribution was as follows:

Airframe with 4 x BMW 801 TG engines fully equipped, without turbo-jet units and fuel	26,500 kg
2 x BMW 018 turbo-jet units	5,000 kg
Fuel	20,000 kg
Lubricants	2,000 kg
Take-off weight	53,500 kg

Without the turbo-jets engaged, a maximum range of 10,500 km could be expected with throttled engines, at an altitude of 3,000 m and at an average speed of 250 km/h. Maximum operational speed with an average load was estimated at 530 km/h. Once the turbo-jets had been engaged at full thrust, maximum range would be 4,500 km at a speed of 666 km/h including a climb to 12,800 metres. At a height of 7,000 metres, maximum speed would be 792 km/h.

According to von Stotzingen and Huber: '*Consequently, this aircraft permits sorties with intermediate high-speed flying, at high-altitudes within the mentioned ranges. This means this aircraft is well-suited for all the missions in which long ranges and long flight durations, combined with temporary high speed flying to out-distance the fighter defence, are required, such as for long-distance reconnaissance flights over enemy territory or for supporting submarine operations over water where there is enemy air superiority.*'

Their report concluded that for operations: '... *the proposed variant of the Me 264, despite its great range and flight duration which, as previously mentioned, could be increased by the use of auxiliary tanks, has a considerable speed reserve which decreases the range only if it is utilised for an extended period of time. This high-speed potential enables the aircraft to out-distance enemy fighter defence, especially over water, since, even if the enemy has any jet fighters, the ranges of these fighters will be somewhat limited. This means an aircraft of this type has a number of important technical advantages for supporting submarine operations and can be used for various missions such as ultra-long-range reconnaissance operations, or at lesser ranges, for combat missions.*'

'Such missions could be accomplished on the following basis: take-off undertaken at full power and with turbojets on full thrust. Taxiing distance approximately 1,500 metres with a fuel load of 20 tons. The aircraft would then climb to 4, 000 metres at which point the turbojets would be stopped. The aircraft would then continue to fly for a distance of 2,330 km at 3,000 metres altitude at optimum engine setting. Before reaching the target, the aircraft then climbs with turbojets engaged to 12,000 metres and flies at a high speed of 758 km/h at this altitude for between one and one-and-a-half hours. The aircraft then returns to 6,000 metres altitude with optimum engine setting. Before reaching the home field, it again flies for half an hour at 10,000 metres at 780 km/h in order to elude any possible fighter attack. Landing is performed with 1,000 kg fuel reserve following a total flying time of 14.25 hours.

'In conclusion it may be stated that the combined piston engine and turbojet drive, in an aircraft of a suitable type, offers possibilities which are unattainable by pure piston engines or pure turbo-jet propulsion and offers new possibilities for combat use.'

As an alternative form of power and propulsion, studies were also made of the potential to develop a regenerative heat exchanger engine for the Me 264 by *Dr* Ludolf Ritz, a de-icing specialist at the *Aerodynamische Versuchsanstalt* (AVA – Aerodynamic Test Establishment) at Göttingen. According to recent research conducted by Antony L Kay into German gas turbine development between 1930-1945[i], these studies centred around the desire to develop heat exchangers which, by feeding back to the combustion

Below The Brown Boveri & Cie turboprop project using twin Ritz heat exchangers as proposed for the Me 264.

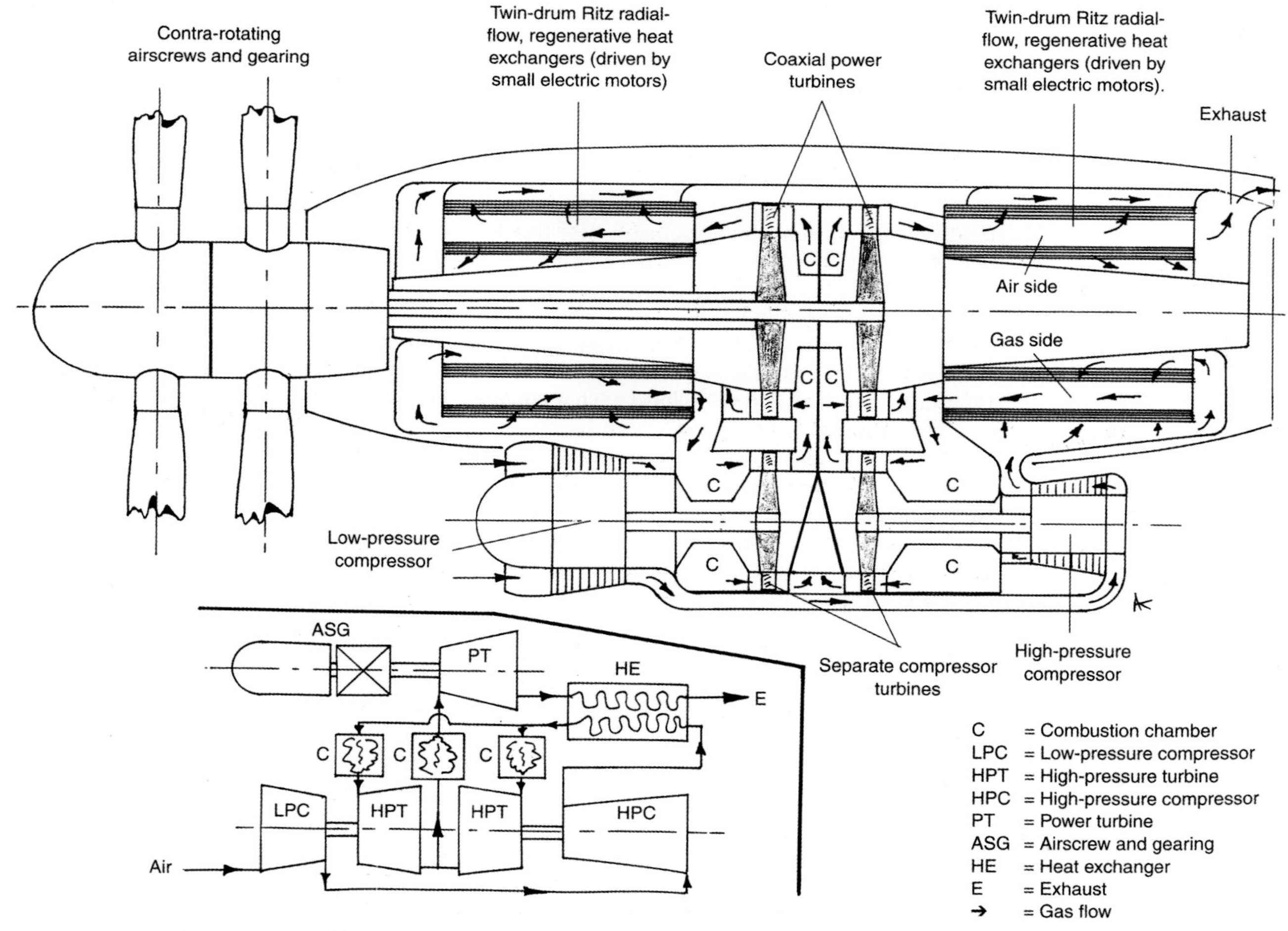

i See *German Jet Engine and Gas Turbine Development 1930-1945*, Antony L. Kay, Airlife Publishing, Shrewsbury, 2002, pgs 220-224

Right Plan and front views of the Me 264 fitted with the proposed Ritz heat exchanger propeller turbines of around 5,000 shp each. Because of the improved fuel economy it was calculated that the Me 264's range would have been doubled. Either Brown Boveri or AEG engines were to be fitted and this drawing shows the latter option.

chamber some of the heat normally wasted in gas turbine exhaust, could reduce fuel consumption. Unlike a recuperative heat exchanger (in which the hot and cold streams are divided by thin metal surfaces such as tubes), a regenerative type exposes a suitable material such as porous refractory, alternately to the hot and cold gas streams, picks up heat from one and then gives it up to the other.

Ritz and his team of eight assistants embarked upon the challenge of designing a regenerative heat exchanger which would both improve gas turbine efficiency but be light enough and small enough to use with aircraft engines. In addition to aircraft gas turbines, other usages were proposed such as closed-cycle gas turbines for fast boats and torpedoes. Theoretically, Ritz's heat exchanger would offer a significant improvement in gas turbine performance.

After investigating the alternative of either an axial flow or a cross-flow type of heat exchanger, it was decided that the latter type was more suitable and efficient. The design incorporated a rotating cylinder or tube constructed from multi layers of woven mesh which gave local turbulent motions to the flow, thereby increasing heat transfer with little effect on pressure loss. Comparison studies were made between the Me 262 and the Me 264. These indicated that while no heat exchanger should be used with the jet fighter, heat exchange used with the Me 264 fitted with two turboprop engines doubled range owing to the substantial reduction in fuel consumption.

Acting on this study, Brown Boveri & Cie (BBC) in Heidelberg and AEG in Berlin were requested by the *Technisches Amt* to design a turboprop engine incorporating twin Ritz regenerative heat exchangers for the Me 264. BBC proposed a design using two compressors, each driven by its own turbine, with two power turbines to drive the contra-rotating airscrews, with reheat taking place at one stage. The compressors and their turbines were mounted below the heat exchangers, which were coaxial with the airscrew turbines and were rotated by small electric motors. Fuel consumption was estimated at 140.7 gm/thrust hp/h compared to 207.9 gm and 181.6 gm in the piston and diesel engine respectively as fitted to the Me 264.

The AEG design was similar to the BBC proposal, but was arranged horizontally with the wing instead of being partly underslung beneath it.

By the end of the war however, little had progressed beyond the drawing board with only laboratory models constructed for testing.

In theory, this promised a phenomenal improvement in gas turbine efficiency, but by the end of the war only laboratory models of aircraft heat exchangers had been tested (with encouraging results) and various projects had been drawn up.[43]

Finally, on 11 August 1943, the Me 264 V 1 was taken out of service at Lechfeld, its Jumo 211J-1 engines were removed and it was re-equipped with BMW 801 MG/2 twin row radial engines. The aircraft also assumed the internal Messerschmitt designation 'M IV'.[44]

Chapter Five

'With just one or two bombs...'

'This Me 264, with its combination of piston engines and turbo-jets, offers us prospects of revolutionary importance...'

***Hauptdienstleiter Dipl.-Ing.* Karl-Otto Saur, June 1944**

On 21 August 1943, the OKM expressed its annoyance to Milch over Göring's latest order to stop the production of seaplanes in favour of fighters for the intensifying war against the Allied bombers now appearing in ever greater numbers over the Reich. Besides a need for the Bv 222 as an interim measure before the arrival of the Ju 290 long-range reconnaissance aircraft, the OKM reminded Milch of the important need for long-range operational machines.[1]

Meanwhile at Augsburg, Karl Seifert was grappling with specifications and performance tables associated with the Me 264's range and fuel consumption; there were also the possibilities of jettisonable fuel tanks, and the fitting of extra tanks to the wings. Seifert was proposing to fit the '*Sudeten*' with faired external tanks of 4,000 and 6,600 litres capacity for ultra-long-range missions using the 56-ton BMW 801 G and E-engined *Fernaufklärer* and *Sonderfernaufklärer* variants.[2] The tanks, which were divided into three separate compartments, were to carry C3 fuel and in the case of the 6,600 litre tank was designed to hold a reserve capacity of 250 kgs for the *Sonderfernaufklärer* and 470 kg for the *Fernaufklärer*. Plans were made to fit a small tail unit to the 4,000 litre tank so as to allow it to 'glide' away safely from the aircraft after release. It was proposed to carry out wind-tunnel testing on models of the wing and tanks to assess the aerodynamics involved and also the centre of gravity. Of great concern however, was the possibility of emergency release during take-off, when using such large tanks. As far as is known, however, nothing further was progressed on this and no models were made.

On 14 October 1943, the USAAF VIII Bomber Command despatched 291 B-17s on a second mission to aircraft industry plants at Schweinfurt in southern Germany, targets which had proved so costly to the Americans in August. The air war over Germany was about to escalate to new heights. A massive air battle ensued, at the end of which the *Jagdwaffe* had lost 31 of its fighters destroyed, 12 written off and 34 damaged – between 3.4-4 per cent of available fighter strength in the West. The Americans lost 60 B-17s

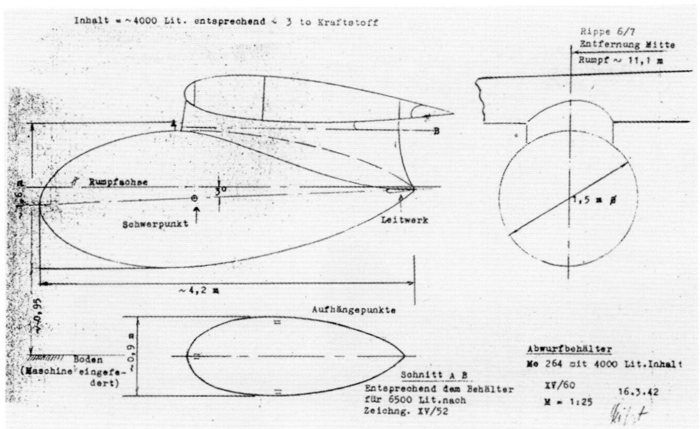

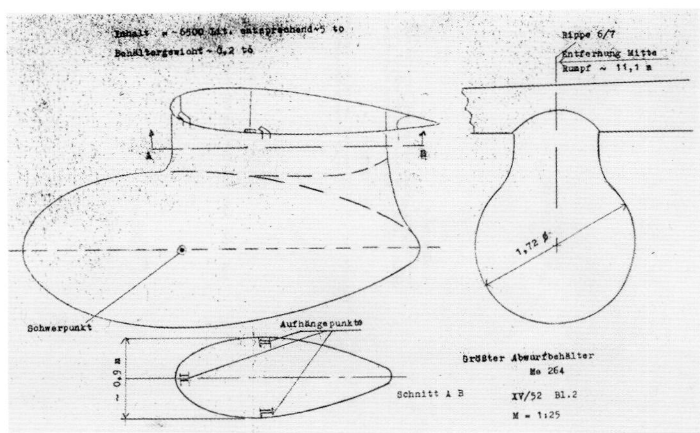

Above and above right
Messerschmitt drawings showing the proposed vast 4,000 litre and 6,500 litre faired external drop tanks planned for the Me 264 when undertaking ultra-long-range missions.

and 600 aircrew during the raid. Seventeen more bombers were seriously damaged and a further 121 were damaged but repairable.

As the bombs fell and the *Luftwaffe* did battle with the Americans, to the north, in Berlin, Göring met with Milch, Friebel, Petersen, Peltz and Messerschmitt to discuss aircraft development possibilities. It was reported that the Ta 400 project from Focke-Wulf was suffering from a lack of experienced designers and assembly workers. Messerschmitt unhesitatingly used the moment to promote the Me 264 and stated his belief that his competitor for a transatlantic reconnaissance aircraft would find it difficult to produce an aircraft with a range of just 10,000 km, whereas the Me 264 could soon be available for reconnaissance missions of up to 14,000 km or 18,000 km. Surely his Messerschmitt would be more acceptable given the existing policy of not developing two machines of the same type simultaneously?

According to Messerschmitt, the components for the first five prototypes were completed, but he lacked the necessary space and facilities in which to construct them. In order to make space available for production of the Me 410 heavy fighter, all the Me 264 final assembly building jigs had been moved from Augsburg and placed in storage at Gersthofen.

Göring knew what Messerschmitt was asking for, but he also knew that the availability of an entire production facility for a long-range bomber – including the necessary workforce – was a pipe dream and therefore a significant number of such aircraft could not realistically be expected, irrespective of whether it was to be the Ta 400, Ju 390 or Me 264. Considerably more promising to Göring was the Junkers production facility at Prague. In any case, the Focke-Wulf project was out of the question because its development in France was way behind schedule.

Messerschmitt would not give up; he explained to the *Reichsmarschall* that the Ju 390 which he favoured, could never – at least in terms of performance – become the ideal machine for long-range operations over the Atlantic. The inadequate engines were still presenting a problem and thus the Me 264 would prove of greater development potential than the Ju 390.

Not surprisingly with Milch and Messerschmitt facing each other across the table, the meeting soon became heated. It was announced that in the view of Ernst Heinkel only the Ta 400 and the reworked He 177 would be capable of effectively undertaking long-range bombing, anti-shipping or reconnaissance missions. Heinkel apparently considered that the Me 264, by comparison, was purely '...a record-breaking aircraft which does not come up to the service requirements for operations in large numbers.'

Göring's voice boomed across the room. He wanted to discuss offensive air operations against the United States and asked *Oberst* Petersen what the shortest distance was from German-held territory to New York. Petersen replied that from Brest it was 5,400 km.

'If only we could reach it!' the *Reichsmarschall* erupted, 'With just one or two bombs we could force them to blackout!'

Peltz then enquired as to the bomb load that the Me 264 could carry and Friebel replied that an ideal configuration would be two 3,000 kg bombs for a distance of 7,000 km.

Eventually Milch proposed stopping work on the Me 264 altogether, in order to put effort in to the Me 262, to which Göring agreed. The following day, production of the Focke-Wulf Ta 400 was cancelled, primarily because Focke-Wulf output was required for production of the Fw 190 D-9 and Ta 152.[3]

On 29 October 1943 Milch angrily told *Professor* Messerschmitt that: 'The Me 264 will not win the war but the Me 262 can. We must therefore concentrate all our efforts on this machine. When Messerschmitt has completed this order and has no other important tasks to deal with, I would then like to solve the question of the Me 264 once and for all. What gain should I get from a machine with a promised range of 20,000 km when it breaks to pieces during take-off, even if it happens to only ten per cent of them?'[4]

A week earlier, on 22 October, *Flugkapitän* Vogel from Department E4 of the *Erprobungsstelle* Rechlin arrived at Augsburg to discuss navigational refinements to the Me 264 *Sonderfernaufklärer* with Seifert, Konrad and other members of the *Projektbüro*. In the light of Milch's decree of the 14th, a discussion on such detail seems strange, but nevertheless, specifications and dimensions were agreed and/or revised in relation to wiring, the height of the navigator's table and storage for apparatus, equipment, charts and books.[5]

By 12 November, Messerschmitt had issued a more complete project specification for a *Sonderfernaufklärer* in accordance with the requirements of the *Versuchsstelle für Höhenflüge.*[6] This contained various refinements such as illumination for the navigation room during night flying, retractable antennae, Flak and emergency recognition pistols and flares, ultra-violet lighting for the instrument panels, the installation of a magnetic compass, a photo counting calculator to be fitted behind the right-hand side of the cockpit, footrests for the B-1 gun position, curtains in the cockpit and heating for the lubrication in the fuselage.

By early 1944, the Messerschmitt Me 264V 1 was still grounded at Lechfeld undergoing engine change and further improvements, but construction work on the second and third prototype machines – the V 2 and V 3 – were at a quite advanced stage and progressing well, in readiness for fitting with BMW jet engines.

Due to the earlier experiences with the V 1 flight tests however, studies were still being made on the problems of vibration and shimmying which affected the installation of the nosewheel. On 13 January 1944, *Dr.-Ing.* E. Maier of the *Forschungsinstitut für Kraftfahrwesen und Fahrzeugmotoren* (Research Institute for Motor Vehicle Technology and Vehicle Engines) at the *Technisches Hochschule* (Technical College) in Stuttgart, published a report following trials designed to assess the differences in flutter between a nosewheel fitted with forked suspension struts and one without.[7] However, no firm conclusions were reached and Maier suggested making further experiments. Eventually, friction dampers were replaced by hydraulic dampers and the wheel fork bent backwards in order to cure the problem.[8]

Possibly influenced by the von Stotzingen and Huber proposals from October 1943, by early 1944, the *Projektbüro* was at work on an enhanced, long-range bomber version of the Me 264 designated the P 1085, which incorporated swept-back wings, four piston engines and two additional jet or turboprop engines. The aircraft was to be armed by four MG 151 Zs housed in turrets. The design actually formed the basis for a number of subsequent Me 264 projects for which it was proposed to use only jet engines. The reality however was that such proposals were never likely to get beyond the drawing boards due to the lack of production capacity, especially when it came to engines.

Left and below: In late 1943 tests were conducted by the Forschungsinstitut für Kraftfahrwesen und Fahrzeugmotoren at the Technisches Hochschule in Stuttgart to assess the differences in flutter between an Me 264 nosewheel fitted with forked suspension struts and one without. These photographs show the forked nosewheel rigged into a specially constructed trailer from which flutter data could be obtained. Note the difference in numbers of weight blocks in the four corner casings used to replicate weight and pressure. The size of the nosewheel is almost identical to that of the wheels on the lorry. The sign on the rear of the lorry warns 'Caution! Brake Testing'.

With the turn of the year, various meetings continued to take place on both the future development of the Me 264 and how best it could be deployed – when, and if, it appeared. On 11 January, a long meeting was held in Vienna between representatives of the RLM and the Messerschmitt and Heinkel companies. Heinkel had come up with the novel idea of creating a 56,000 kg 'hybrid' long-range aircraft with a range of 12,500 km by proposing to fit the wings of the Me 264 to the fuselage of the He 277. This was Heinkel's latest long-range/heavy bomber project which was based on the He 177 fuselage, with additional tail armament and – unlike the He 177 – four separately mounted BMW 801 E engines. However, on paper, Focke-Wulf's Ta 400 still had the edge in terms of range (by some 1,500 km) over the He 277.

Right The four-engined Heinkel He 277 V 1 was converted from a He 177 A-3/R 2 and fitted with four separate DB 603 A engines as opposed to the DB 606 as found on the He 177. Heinkel proposed the creation of a 56,000 kg 'hybrid' long-range aircraft with a range of 12,500 km using the wings of the Me 264 mated to the fuselage of the He 277.

The immediate benefit of mating the Me 264 with the He 277 lay in the latter's fuselage design which would, in theory, accommodate more fuel and, at 3,000 kg, a greater bomb load than the Messerschmitt. Furthermore, it was believed that such a mating of the designs would allow heavier defensive armament in the form of quadruple-barrel nose and tail gun positions. However because of its weight, take-off of a fully-laden aircraft could only take place using RATO units and a jettisonable undercarriage. Work on the project continued until the spring of 1944, by which time calculations showed that initial range projections had been over-optimistic and only some 8,000 km could be achieved.

It will be recalled from the previous chapter that there seems to have been considerable variations in opinion as to just how many Me 264 prototypes were actually planned and authorised. Certainly, as far as Milch was concerned – and thus the RLM – three such aircraft had been sanctioned; Willy Messerschmitt appears to have assumed that he had authorisation to proceed with five or six and it will be recalled that *Oberstleutnant* Ulrich Diesing of the RLM *Technisches Amt* referred to six prototypes at a meeting the previous July.

However, on 25 January *Oberleutnant* Wolfgang Nebel of 2./*Versuchsverband Ob.d.L* travelled to Augsburg to meet with Paul Konrad, Karl Seifert, Julius Reicherter and a *Herr* Hugo of Messerschmitt A.G. to discuss the building of five Me 264 prototypes. It seems that *Oberleutnant* Nebel had served earlier with *Oberstleutnant* Theodor Rowehl in the *Versuchsstelle für Höhenflug* with whom amongst other tasks, he had flown a number of secret missions dropping spies and saboteurs over Iraq, Iran and other areas of the Middle East.[9] Later he was assigned from 2./*Versuchsverband Ob.d.L* to the *Stab* of GL/C-E (the *Amtsgruppe Entwicklung*, or Development Department) under Pasewaldt at the RLM for the purpose of assessing the development of new aircraft, particularly long-range types.

By this stage, the plan to mate the Me 264 with the He 277 had been abandoned, and it had been decided to proceed with the construction of the five prototype Me 264s at Augsburg using the plant's available materials and capacity, such as it was. It had further been decided to no longer equip the V 2 and V 3 machines with jet engines but Messerschmitt had been requested by the RLM to conduct a feasibility study in to installing BMW jet engines into the planned V 4 and V 5.

From the surviving minutes of the meeting between Nebel and the Messerschmitt engineers,

Below Oberleutnant Wolfgang Nebel (second from left) talks with Oberleutnant Horst Götz of the Versuchsverband Ob.d.L. (Experimental Unit of the Supreme Commander of the Luftwaffe). Nebel had flown a number of spy-dropping missions over the Middle East, before becoming involved in long-range aircraft development for the RLM. In late 1943, he was assigned to liaise with Messerschmitt A.G. on the development of the Me 264 and would later assume command of Sonderkommando Nebel which was charged with deciding excactly how best to deploy the 'Amerika Bomber'.

much ground seems to have been covered.[10] A 'non-obligatory preliminary plan' for the first three prototypes was presented to Nebel, together with an outline of the proposed armament which was at least equal to or even superior to that of the He 177. Hangar space would be made available by Messerschmitt at Augsburg for the construction of the V 2 and V 3 but, in view of the capacity required, not – for the time being – for the V 4 and V 5. In terms of design staff, it was envisaged that the Me 264 programme would require 50 experienced designers, project engineers and managerial staff. For final series development, Messerschmitt estimated the requirement at 150 personnel, though a further four staff would be required if – as suggested by the RLM – French designers were to be employed. This option was to be discussed at a meeting planned in Paris for 31 January 1944 at which *Oberleutnant* Nebel, *Dr.* Burkhardt of GL/C-E II of the RLM and *Herr* Bley from Messerschmitt A.G. Augsburg would be present.

For the workforce Messerschmitt projected that 500 'skilled workers' would be needed for the construction of the five aircraft; the minutes noted: '*It has to be stressed that the labourers requested must largely consist of truly skilled workmen, since the prototypes have to be built with design drawings (partially) not intended for series production.* Oblt. *Nebel will take over the hiring of workmen in cooperation with the RLM, Messerschmitt A.G. having to give all possible assistance.*'

The meeting concluded with the Messerschmitt representatives emphasising to Nebel that design and construction work could only commence once the required labour force had been made available. By 3 February, the Messerschmitt representative office in Paris signalled *Herr* Bley at Augsburg: '*Dr. Burckhardt and Herr Scheibe* [i] *will be in Paris from 3.2.44. Negotiations regarding Caudron drawing office staff and designers for Me 264. Presence of Herr Bley requested from 4.2.44.*'[11]

As is evident from surviving Messerschmitt records, Karl Seifert continued to struggle with the issue of range and ever-increasing weight. On 10 February, he prepared a revised specification and weight assessment for the '*Sonderaufklärer*' variant of the Me 264. This variant was to be powered by four BMW 801 G engines, with GM 1 nitrous oxide power boost for 25 minutes duration, to carry three RB 50/30 cameras and to be able to accommodate 12 troops or passengers. Armament was to consist of one Arado-built FPL 151 gun turret with a PVE 6 periscopic sight and an MG 131 in a DL (*Drehlafette* – rotating gun mount) 131 turret.

Range was projected at 13,000 km with small external tanks and up to 15,200 km with larger auxiliary tanks. Flight duration was calculated at 42 hours or 47 hours depending on the type of external tank carried, with speed at 6,000 metres being 410 km/h. The aircraft would require a take-off distance of up to 3,600 metres without RATO units, or 2,050 metres with RATO units fitted.[12]

His calculations were as follows[13]:

Wing assembly	5,287 kg
Fuselage	2,164 kg
Tail assembly	421 kg
Controls	142 kg
Landing gear	1,424 kg
Total:	**9,438 kg**

Engines (BMW 801 TC), propellers, GM-1 power boost, and associated equipment:		10,103 kg
Ancillary equipment, engine controls, navigational equipment, hydraulics etc		1,614 kg
Armament (with ammunition)	749 kg	
Camera equipment	247 kg	
Reserve	500 kg	
Total accumulated weight:	**22,651 kg**	

Seifert then calculated three overall weight options allowing for different fuel loads:

A) Without external tanks

Crew	540 kg
Supplies	100 kg
Emergency lubricant	288 kg
GM Boost	697 kg
Tanks protected	13,680 kg
Tanks unprotected	5620 kg
Fuel in external tanks	-
Lubricant	432 kg
Plus Total accumulated weight	22,651 kg
Total weight:	**44,008 kg**

i This is probably *Flugbaumeister* Scheibe who would test-fly the Me 264 V 1 in April 1944 .

B) With jettisonable undercarriage (total four wheels, two fixed, two jettisonable)

Crew	540 kg
Supplies	100 kg
Emergency lubricant	288 kg
GM Boost	697 kg
Tanks protected	13,680 kg
Tanks unprotected	5,800 kg
Fuel in external tanks	4,362 kg
Lubricant	432 kg
External tank weight	450 kg
Plus Total accumulated weight	22,651 kg
Total weight:	**49,000 kg**

C) With heavier jettisonable undercarriage
(total six wheels, two fixed, four jettisonable)

Crew	540 kg
Supplies	100 kg
Emergency lubricant	310 kg
GM Boost	697 kg
Tanks protected	13,680 kg
Tanks unprotected	5,800 kg
Fuel in external tanks	11,082 kg
Lubricant	640 kg
External tank weight	500 kg
Plus Total accumulated weight	22,651 kg
Total weight:	**56,000 kg**

During March 1944, the Allied air forces inflicted damage on the Me 264 programme and also gained vital information on the very existence of the '*Sudeten*'. On the 18th, B-17 Flying Fortresses of the USAAF Eighth Air Force's 1st Bomb Division attacked a snow-covered Lechfeld as part of a concerted strike against a number of airfields in southern Germany. The Me 264 V 1 was slightly damaged as a result of the raid, but it was repaired quickly. As far as is known, the uncompleted V 2 and V 3 airframes were undamaged.[14] However, at some stage after the attack, it was decided to move the still uncompleted V 2 and V 3 north to the so-called '*Metallbau Offingen*', a small production facility in the village of Offingen near Günzburg, to the east of Neu-Ulm, where bomb bays and armament were to be installed. The move was also made to alleviate the increasing production bottlenecks affecting construction of the Me 262.

There were plans to build the V 2 with temporary armament consisting of one MG 131, three MG 151 Z and several lateral window mounts. The weight of such an aircraft with this armament was estimated at 50,000 kg, and the OKL believed that fitted with BMW 801 engines, the aircraft would be capable of a cruising speed of 350 km/h and a range of 9,500 km. Based on such a specification, the Messerschmitt *Projektbüro* produced numerous variations to the basic plan. Three revised series versions of the Me 264 emerged:

Version A: Long-range reconnaissance
Range: 13,600 km (with two auxiliary tanks).
Maximum speed at 6,300 m = 580 km/h.
Max flight duration = 40 hours.
Three cameras to be fitted at rear.

Version B : Long-range bomber
Planned with four BMW 801 E engines and two further Jumo 004 C jet engines. Defensive armament was to consist of an MG 131 in the A and B turret positions, an HD 151/Z in the B-2 position and an MG 131 in C turret position. Two MG 131s were planned for the waist positions.

Gross weight between 48,100 and 49,900 kg depending on whether the aircraft was fitted with the two jet engines.

The range with a 3,000 kg payload was 11,600 km without jet engines and 8,500 km with jets. Calculated maximum speed at 6,400 m would have been approximately 577 km/h, while with Jumo jet engines fitted, approximately 655 km/h at 6,700 m. Due to its pressurised cabin the aircraft would be able to operate at altitudes up to 14,500 m.

The Me 264 B was intended for long-range anti-shipping operations. As with the long-range reconnaissance version it was to be equipped with four Jumo 222 E/F high-altitude engines and two additional jet engines. Its maximum offensive load was to consist of six SCX/SD 1000 bombs.

The full-vision cockpit was to be replaced by a stepped version, similar to that proposed for the Ta 400, which would be less vulnerable from enemy fire.

The defensive armament for the Me 264 B was revised on several occasions up to August 1944, but finally settled upon 360-degree revolving turrets to be equipped with two MG 213s.

Version C : Special long-range reconnaissance ('*Sonderaufklärer*')
Provisionally to be fitted wih a pressurised cockpit, but not confirmed. This version was to carry three Rb 50/30 automatic cameras and defensive armament was to consist of an MG 131 in the A and B turret positions, an HD 151/Z in the B-2 position and an MG 151 in C turret position. Two MG 131s were planned for the waist positions.

Two drop tanks would allow the machine a range of 13,600 km and a top speed of almost 580 km/h at 6,300 m. Maximum projected flight duration was 41 hours. Messerschmitt planned a further variant of this version featuring two additional Jumo 004 jet engines or BMW 801 E/F high-altitude engines and submitted plans to the *Luftwaffe* ordnance specialists for evaluation.

A long-range transporter able to carry 12 to 17 paratroops armed with one FHL (*Fernbedienbare Hecklafette* – remotely-controlled rear gun mount) 151/Z was also planned, but no detailed designs were submitted.

From the V 4 it was planned that all further prototypes as well as first series aircraft should be fitted with four high-performance BMW 801 E engines with turbo-charger and GM 1 system. The GM 1 tanks were to be installed in the centre section of the fuselage.

It is clear that as early as 26 December 1943, the Allies were aware of the Messerschmitt Me 264, since on that date, following information gleaned from captured *Luftwaffe* personnel, a British Air Intelligence officer wrote that: '... *Me 264 is of a rather spectacular nature... particular emphasis is laid upon range and is sufficient to attack the USA.*'[15]

Then, on 28 March 1944, the Allies were offered an enigmatic picture of the Me 264 – the first real proof of the existence of the '*Amerika Bomber*' – when a Mosquito IX of 540 Squadron, RAF, photographed Lechfeld airfield.

The 'Lechfeld 127'

At 1325 hours on 28 March 1944, a Mosquito PR IX, LR435, piloted by F/Lt A.T.Leaning of 540 Squadron, RAF, together with his navigator, F/O S.G. Dale, took off from Benson airfield in Oxfordshire and flew towards Beachy Head on the South Coast of England from where it made course east across the English Channel. Leaning and Dale had been briefed to conduct a routine photo-reconnaissance mission of airfields in southern Germany.

F/Lt Leaning later recorded: '*We crossed the coast north of Dieppe on course for Liege where we turned for Stuttgart. Photos taken of Stuttgart, Ulm, Langweid airfield, Augsburg airfield, Lechfeld airfield, Penzing airfield, Weazling airfield, Munich, Landshut and airfield, Schweinfurt airfield.*'[16]

As Leaning and Dale passed over Lechfeld, they happened to catch the Me 264 on film with their F52 high-altitude camera. The aircraft was in the open, on grass, not far from the main hangar just off one of the airfield's concrete taxi tracks just at the time it was undergoing lengthy engine change and maintenance.

Having successfully accomplished its mission, the lone Mosquito turned for home and made a safe return to Benson, landing at 18.10 hours. The cameras were unloaded and the film sent to the RAF's Central Interpretation Unit (CIU) at Medmenham in Buckinghamshire for analysis.

On 10 April, the CIU issued an extremely accurate report on what it termed an 'Unidentified four-engined aircraft – possibly of Messerschmitt design, "Lechfeld 127"'.[17] This was the first time the Me 264 had been seen by the Allies since F/O A. Stuart and F/Sgt M. Pike, also of 540 Squadron, had photographed a 'very large aircraft' at Lechfeld on 10 March 1943 (see Chapter Four). Medmenham concluded that the aircraft, which was '*very indistinctly visible*', '*...can now be identified as a "Lechfeld 127".*'

The report continued: '*Among the aircraft seen recently at Lechfeld, near the hangar believed to be used by Messerschmitt for developmental work, was a very large four-engined aircraft, probably with a tricycle undercarriage. This aircraft, which is seen with a Do 217, will be known for the present as the "Lechfeld 127".*

'Design characteristics
Four-engined mid-wing (or possibly shoulder-wing) monoplane, with a wing of fairly high aspect ratio. The leading edge of the wing is swept back, and the trailing edge appears to be straight. The tail unit is high off the ground, indicating the likelihood that a tricycle undercarriage is fitted. The nose is not long, and the fuselage tapers towards the tail unit, which is set high and has twin fins and rudders. The tailplane has a swept back leading edge and a fair amount of dihedral. The engine nacelles are staggered and appear to be centrally mounted.'

'Unidentified four-engined aircraft – possibly of Messerschmitt design ... very indistinctly visible ... can now be identified as a "Lechfeld 127"' – was how the RAF's Central Interpretation Unit summarised the Me 264 from the photograph obtained by F/Lt Leaning and F/O Dale following their flight over Lechfeld on 28 March 1944.

Above On 14 April 1944, the Me 264 V 1 was fitted with four new BMW 801 MG/2 TC-1 air-cooled 14-cylinder radial engines.

Below The Me 264 development team take a break in the open air during flight-testing with the new BMW 801 engines at Lechfeld or Memmingen in early 1944. Note the mock-up MG 131 Z gun turret aft of the cockpit and the open window panel allowing some much needed ventilation into the glazed 'greenhouse'. Karl Baur is standing to the far left.

On 14 April with four new BMW 801 MG/2TC-1 air-cooled 14-cylinder radial engines finally fitted, certain technical and design enhancements incorporated and its damage repaired from the American bombing raid the previous month, the Me 264 V 1 emerged from the hangar at Lechfeld to recommence its test programme.

In the first instance, Karl Baur ran up the engines for just over four hours, before embarking on a series of taxiing runs. Unfortunately, during these tests, the brake shoes tore off and the aircraft had to return to the hangar for repairs. It was an inauspicious start.[18]

Following the completion of repairs, the aircraft was once more declared fit for testing and on 16 April, in what was to be a busy day for the aircraft, it made no fewer than six flights. Taking to the air during the morning for the first time in eight months – its 37th flight – for 24 minutes, Karl Baur reported no real problems, but he did discover that the engine and propeller synchronisation was out and that fuel pressure was too great. In general terms however, the aircraft performed well and Baur opined that the improvement offered to the machine by the fitting of the BMW engines was 'magnificent... especially a much better performance on take-off.'[19]

There was now a *Luftwaffe* presence in the test programme, in the form of *Oberstleutnant* Siegfried Knemeyer, the new head of the RLM *Amtsgruppe Entwicklung* (GL/C-E), and *Flugbaumeister* Scheibe who had arrived at Lechfeld to inspect the Me 264 and to join in the flight-testing. Prior to the war, *Oberstleutnant* Knemeyer had worked on the development of instrument flying and had qualified as an *Ingenieur*. He later served with the *Aufklärungsgruppe Ob.d.L.* under Rowehl, flying test flights and operational missions in many different aircraft over Europe and North Africa. In April 1943, he was appointed Technical Officer to the *Angriffsführer England*, the command responsible for bombing raids against the British Isles and was then assigned to Göring's personal staff where he applied pressure to encourage development of the Me 262.[20] As a skilled pilot, engineer and with a considerable knowledge of electronics, his involvement in the Me 264 programme would be valued. It was Knemeyer who dubbed the Me 264 '*das Schiff*' – 'the ship'.[21]

Meanwhile, in the light of continuing Allied bombing raids (Augsburg had been targeted by the US Eighth Air Force on 13 April), the decision was made to move the Me 264 from Lechfeld, further south to the large airfield at Memmingen. It flew the short distance to its new base in 12 minutes, where its emergency skid was torn away as a result of touching down too heavily during landing. It is probable that Knemeyer and Scheibe accompanied Baur on this flight.

The aircraft then returned to Lechfeld where, presumably, the skid mount was repaired, before *Flugbaumeister* Scheibe took the controls for a 35-minute flight. Having deliberately stopped two engines in flight, Scheibe considered the aircraft to be generally very good technically and that when the rudders were fitted with balances, any excessive vibrations – which had plagued the aircraft earlier – almost ceased. Knemeyer then flew '*das Schiff*' for an 18 minute flight. He was favourably impressed by the aircraft's qualities, though he did complain about poor visibility from the cockpit due to reflection which he believed could impede results during a combat mission. He also noted the vibrations in the airframe as the main problem to be addressed, as well as the need for a Patin autopilot system for long-range missions and an emergency compass, which although had been fitted did not yet function. It was also Knemeyer's view that any questions concerning the functioning of the rudder could easily be attended to throughout the course of further testing.[22]

Following its final flight on 14 April, the Me 264 V 1 had amassed a total of 32 hours, 50 minutes flying time since its maiden flight at Augsburg in December 1942.

Above Oberstleutnant Siegfried Knemeyer, the head of the RLM Amtsgruppe Entwicklung, visited Lechfeld in April 1944 to inspect the Me 264 development programme and to test-fly the V 1 prototype with which he was impressed. As a skilled pilot and engineer with a considerable knowledge of electronics, his involvement with the Me 264 would be valued. It was Knemeyer who dubbed the aircraft 'das Schiff' – 'the ship'.

Below The Me 264 V 1 banks in low over either Lechfeld or Memmingen airfield during flight tests with BMW 801 engines in the spring of 1944.

Above The offices of the Messerschmitt Projektbüro at Oberammergau in southern Bavaria occupied the former quarters of a mountain infantry unit. It was here that Oberstleutnant Siegfried Knemeyer and Flugbaumeister Scheibe met with Professor Messerschmitt and his designers and engineers. The meeting concluded that '... the Me 264 must serve as the basis for future large aircraft development.'

After their flights, Knemeyer, accompanied by Scheibe, journeyed south to Oberammergau in southern Bavaria, the new location of the Messerschmitt *Projektbüro* which had moved there together with the company's Construction and Statistics departments to escape the threat of Allied bombing. In a room somewhere within the former mountain infantry barracks, Knemeyer and Scheibe met with *Professor* Messerschmitt and members of the project team to discuss the position with regard to the Me 264. Knemeyer began proceedings by stating that he and *Flugbaumeister* Scheibe were in broad agreement that the programme could be salvaged and continued. Knemeyer, apparently impressed by the way in which Messerschmitt had engineered the aircraft, even went so far as to comment that he thought it '...could be bought straight off the peg'.[23]

Although Knemeyer informed the Messerschmitt team that there were a number of defects which still plagued the aircraft, such as the over-powerful and unharmonized rudder forces, equally he recognised that such defects were largely due to lack of production capacity and were solvable. He also expressed general satisfaction with the overall layout of the aircraft, particularly the weapons installations: '*The weapons installations are very adequate based on existing requirements and should present no problems. The aircraft's internal dimensions make it very suitable for carrying ordnance. The only complaint about the design is the long-known issue of visibility.*'

Scheibe requested more powerful engines – i.e. the Jumo 222 – and was favourable towards the possibility of a six-engined version of the Me 264. However, despite Knemeyer's enthusiastic opinion, that once the aircraft went into series production, it would be regarded as '*the* "ship"', the reality was that the RLM was still uncertain as to which of the Me 264 designs should be continued and which aircraft it should replace.

The minutes of the Oberammergau meeting recorded that it was felt that '... *the Me 264 must serve as the basis for future large aircraft development.*' As a caveat to this however, Knemeyer added that whichever company was eventually selected by the RLM to continue with the development of the Me 264 should be the one to continue testing the V 1; '*For Messerschmitt A.G.*' the minutes recorded, '*this means that it must continue the Me 264 testing until the RLM has made its decision.*'

The next day, flight-testing resumed. Generally, things went well with few problems although from an aerodynamic point of view, it was noticed that the tail felt too heavy during flight. An emergency tail wheel had also been fitted and underwent testing for flutter.

On 26 April, the recently promoted *Hauptmann* Nebel visited Memmingen to fly the Me 264 on an assessment flight for the RLM. It was to be the 43rd test flight and Nebel was airborne for 19 minutes.[24] The aircraft was missing its emergency tailwheel which had been ripped off on an earlier flight.

The following day, flying at a height of 450 metres, the Me 264 reached a speed of 490 km/h at maximum operational setting and 470 km/h flying at maximum continuous power.[25]

Four flight tests totalling just over one and a half hours and conducted between 28 April and 2 May were aimed at solving the vibration problem. In order to avoid the construction of an entirely new tail unit, continued attempts were made to bring the vibrations under control by fitting balance weights. The scheduled performance trials were – once again – delayed. Unfortunately, these efforts proved in vain and the vibrations persisted, despite the weights.

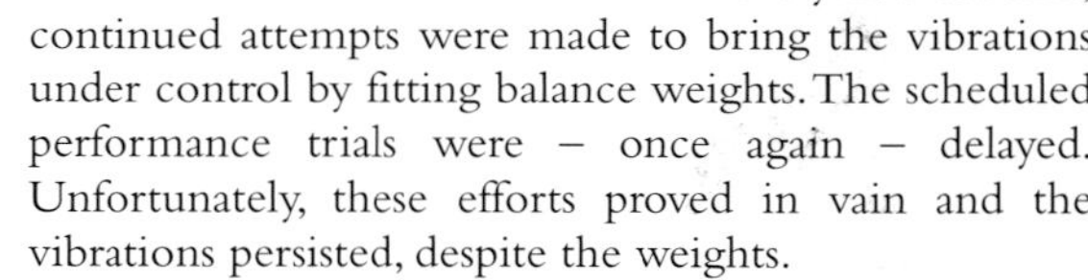

Below Following his test flight in the Me 264 in April 1944, Flugbaumeister Scheibe requested more powerful engines, possibly the Jumo 222.

On 3 May, *Hauptmann* Nebel made three further flights, but still the problem persisted. It now seemed as if a new tail unit was inevitable. Messerschmitt anticipated that wind tunnel tests would provide the required design data by July 1944.[26]

According to *Luftwaffe* historian Manfred Griehl, *Generalmajor* Hans-Henning von Barsewisch, the *General der Aufklärungsflieger*, was also present at Memmingen to fly the '*Sudeten*', but his opinion was not as favourable as Knemeyer's: 'He judged the machine to be too slow for combat missions, though fitted with the BMW 801 MG/2 TC-1 engines it was about 10 per cent faster than with its previous Jumo 211 engines.'[27]

Possibly as a result of Knemeyer's endorsement, Messerschmitt drew up a revised specification for an '*Me 264 – Sonderfernaufklärer*'.[28] There was actually little change over the specification produced by Karl Seifert

in February; camera and armament installation remained the same, but the proposed engine model was to be revised from the BMW 801 G to the 801 E. The key factor however, was that despite a reduced fuel load, range was extended (by 500 km) and flight duration was increased marginally (by between 30 and 90 minutes, depending on final weights). The rate of climb also increased to 4.1 metres per second for a 49-ton specification aircraft and to 2.5 metres per second for a 56-ton aircraft. Maximum service ceiling was increased to 10,300 metres. Using GM 1 nitrous oxide power boost, maximum speed at 8,300 metres was increased to 618 km/h.

Germany's Japanese allies also showed great interest in the Me 264 and throughout 1943 and early 1944 '... a lively interchange of questions and answers' took place between the RLM in Berlin and senior officials within the Imperial Japanese Army Air Force in Tokyo. Additional reports from the Japanese Military Attaché in Berlin included detailed outlines of the Me 264's capabilities together with information on construction and performance. The aircraft was at one time suggested for use in flights between Europe and Asia. In August 1943, Tokyo Army authorities informed Berlin that if the Germans thought the Me 264 could bomb the American mainland, all possible Imperial Army personnel and aid would be given to its development. Two months later, the Military Attaché advised that it would not be proper for Japan to renew requests for the Me 264, since the Messerschmitt firm was struggling to satisfy the more immediate needs of the *Luftwaffe*. Information on the aircraft continued to be sent to Tokyo until March 1944 however, and in May of that year operational was were available for shipment to Japan. While there is no evidence that drawings or parts were ever shipped to the Far East, it is believed that manufacturing drawings were on board the U-boat, U-864, which was sailing to Japan when it was sunk off the coast of Norway by a British submarine in February 1945.[29]

Above In the late spring of 1944, Generalmajor Hans-Henning von Barsewisch, the General der Aufklärungsflieger, deemed the re-engined Me 264 too slow for operations. Messerschmitt subsequently drew up a revised specification, changing the engine type from the BMW 801 G to the 801 E.

Below The fuselage of the 'not finalised!' Me 264 Sonderfernaufklärer prepared on 15 April 1944. This interesting drawing shows the crew seating positions and armament layout, including the FPL 151 gun barbettes in the upper and lower rear fuselage together with the PVE 6 periscopic sight, camera and the DL 131 turret in the upper forward fuselage.

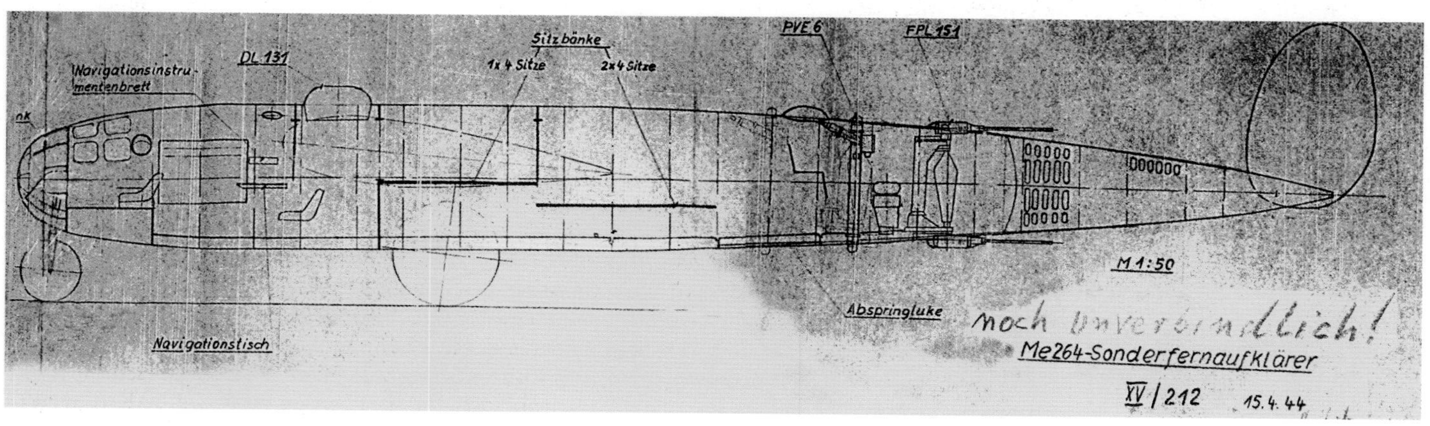

Meanwhile, on Tuesday, 23 May 1944, *Reichsmarschall* Göring was in a bitingly cynical, though remarkably prognostic mood, when he chaired a high-level aircraft production conference attended by the *Reichsminister* for Armaments and War Production, Albert Speer, as well as Milch, the *Luftwaffe* Chief of General Staff, *General der Flieger* Günther Korten, Petersen, Diesing, Knemeyer and Saur at which the He 177's failings were discussed. Göring asked: 'The following point now arises: what can be done to make possible reconnaissance at extreme ranges? What type of aircraft do I have for this purpose? There are the Ju 290 and Ju 390, types which have a colossal fuel consumption, and now – keep a firm hold on yourselves and don't fall under the table – the Me 264! This aircraft, which need be produced only in small numbers and could be further improved, would give us a much faster aircraft than the 290 and would make reconnaissance at extreme ranges possible. I would set a definite limit – and we have all agreed on this point – of one *Gruppe*, which could be set up to strength. This unit would be allotted special long-range reconnaissance duties and we must consider the possibilities if the 264 were especially adapted for this work. If this were done we could exclude all the types which devour so much fuel such as the 290, the 390 and the 288.'[30]

Just when it needed harmony, decisiveness, direction and cooperation, the German aircraft industry and the *Luftwaffe* were suffering from incessant and simmering internal power politics, jealousy, distrust and indecision. The old feud between Milch and Willy Messerschmitt had not gone away. But in May 1944, confusion also reigned following Hitler's question and subsequent requirement that the new Me 262 jet fighter carry bombs. Admittedly, Messerschmitt was under pressure to deliver fighters for the *Reichsverteidigung* and was heavily committed to and distracted by the development of the Me 262. Milch however was publicly against throwing away the jet's speed and air superiority role by turning it into a bomber; yet his star was also on the wane as a result of the *Luftwaffe's* battlefront failures, and as historian David Irving states in his biography of Milch, it was at precisely this time that he '... *recognized that the end of his long road was in sight.*'

At a conference, less than a week later, on the 29th, 'to clarify things once and for all' regarding the deployment of the Me 262, and to which Milch was not invited to attend, Messerschmitt was still openly at loggerheads with the absent State Secretary. The aircraft designer took the opportunity to blame Milch for the prevailing 'misunderstandings' (i.e. delays) which had occurred and which had affected progress on the Me 262.[31]

Far away from the corridors of power, at Memmingen, Karl Baur and the Messerschmitt project team continued the flight programme with more stability tests. On 5 June 1944, Baur once more climbed up the short access ladder, forward of the bomb bay, into the fuselage of the Me 264 and made his way to the glazed cockpit section. After a trouble-free take-off, he spent 40 minutes in the air, at which point the assimilated operational flight at 6,000 m had to be interrupted because lateral control was affected as a result of one fuel tank emptying faster than its neighbour (tank numbers 4 and 5), and there were problems with at least one engine, when the propellers did not reach full revolution. When Baur landed, he also noted that he had experienced repeated malfunctions with the Patin autopilot servos and that this would need urgent improvement.[32] Furthermore, one propeller blade on the left outer engine was badly damaged and had to be replaced having been struck by a stone. The left wing had a hole in it, possibly from earlier bomb damage following an enemy attack on the airfield. The brake cylinder on the left undercarriage was leaking and the internal ballast in the fuselage had to be adjusted. The rebuilding of the counterbalance on the rudder decreased deflection somewhat, but Baur recommended that an enlargement of the fin was still necessary.[33]

The following day, 6 June, the Allies landed in France, pouring 155,000 men plus armour and vehicles onto the Normandy beaches. Fortress Europe had finally been invaded. The Allied air cover was immense with sufficient capability to fly more than 14,500 sorties within the first 24 hours. Slow to react initially, the OKW started to move troops and armoured formations into the Invasion area. The *Luftwaffe* frantically began to transfer some 1,000 fighters to the battlefront from airfields in Germany. Some sources state that Karl Baur undertook a test flight in the Me 264 V 1 on this day, and experienced severe rudder fluttering at speeds between 380-450 km/h, but this cannot be verified.

Whatever the case, Baur now firmly believed that the size of the tailplane was too small and that this was the reason for the Me 264's lack of stability.[34]

As conditions at the battlefronts became increasingly more worrying, so ideas became more desperate. At a meeting of the *Jägerstab* on 10 June 1944 – the committee of aircraft industry chiefs and production specialists set up by Speer and Milch in March and intended to manage and resuscitate the flagging fighter production industry – plans were once again brought up for retaliation against the United States. It was suggested that specially equipped He 177s should be used on one-way missions against North American targets and their crews bale out after attack to become PoWs. The plan did not meet with much enthusiasm.[35]

According to Werner Baumbach, June 1944 was 'the last time' he heard the production of extreme-range bombers discussed, when the tireless and bluntly spoken engineer and expert in armaments production, *Hauptdienstleiter Dipl.-Ing.* Karl-Otto Saur, proclaimed bullishly to the *Jägerstab* that the time for 'botches and bungles' was over. It was time for 'revolutionary inventions.' With this in mind, Saur expressed his whole-hearted support for the continued development of the multi-purpose Me 264: 'I am in favour of going all out with the Me 264. This Me 264, with its combination of piston engines and turbo-jets, offers us prospects of revolutionary importance. Its load and range make it seem adapted for all purposes, even in relatively small numbers.'[36]

But it was less than a month later, on 6 July 1944, that Saur reluctantly told his colleagues Schaaf and Lange: 'The Me 264 will not appear for a very long time. We are attempting to bring it forward by one year. Even then output will be no more than three or four or five at the most, and the aircraft will have to be built by hand so as not to tie down the whole of the plant to this type.'[37]

On 24 June 1944, the V 1 made its 51st test flight for 24 minutes without any perceivable problems. Two days later, Baur took to the air in the Me 264 to make a climb trial flight at operational settings. The aircraft was airborne for 51 minutes, but the flight had to be cut short when the fuel pressure of both inner engines fell to zero. After checking the fuel pumps, several defects were found. Additionally, the Patin autopilot, radio and electrics system all required repairs.[38] From then until 12 July, the flight test programme was put on hold by unseasonably bad weather conditions and a range of minor faults. While the engineers waited for the weather to improve, the faults were rectified and the time used to fit new fuel pumps and to repair the autopilot system.

Towards the end of June, *Oberst* Petersen of the *Kommando der Erprobungsstellen* at Rechlin decreed that, after extensive investigation, neither the Me 264 nor the Ju 390 would be suitable for operations over the Atlantic, since the combined weight of armament, armour, payload and fuel would result in an excessive take-off weight and wing load. Furthermore, neither the strength of the undercarriage designs, nor the existing armament were deemed sufficient. In the case of the latter, it was proposed that turrets should be fitted with 30 mm guns.

At about the same time, the OKM gave its support to Milch's apparent proposal to terminate production of the He 177 only when 'quantities' of the Me 264 entered service – an astonishing about-turn given the *Generalluftzeugmeister's* earlier directives.[39] Quite how – and when – Milch saw the Me 264 being delivered in quantities is not known!

Then, on 9 July, the Me 264 was mentioned briefly at the evening situation conference of the Chief of the *Luftwaffe* Operations Staff, *Generalleutnant* Karl Koller. Following a visit from a naval liaison officer, Koller stated that whilst he recognised there was a need for maritime reconnaissance and that technical problems existed, the Ju 290 and He 177 would be the only aircraft available to fulfil the role in the foreseeable future with the exception of the Me 264 which would only be 'built in small quantities.'[40]

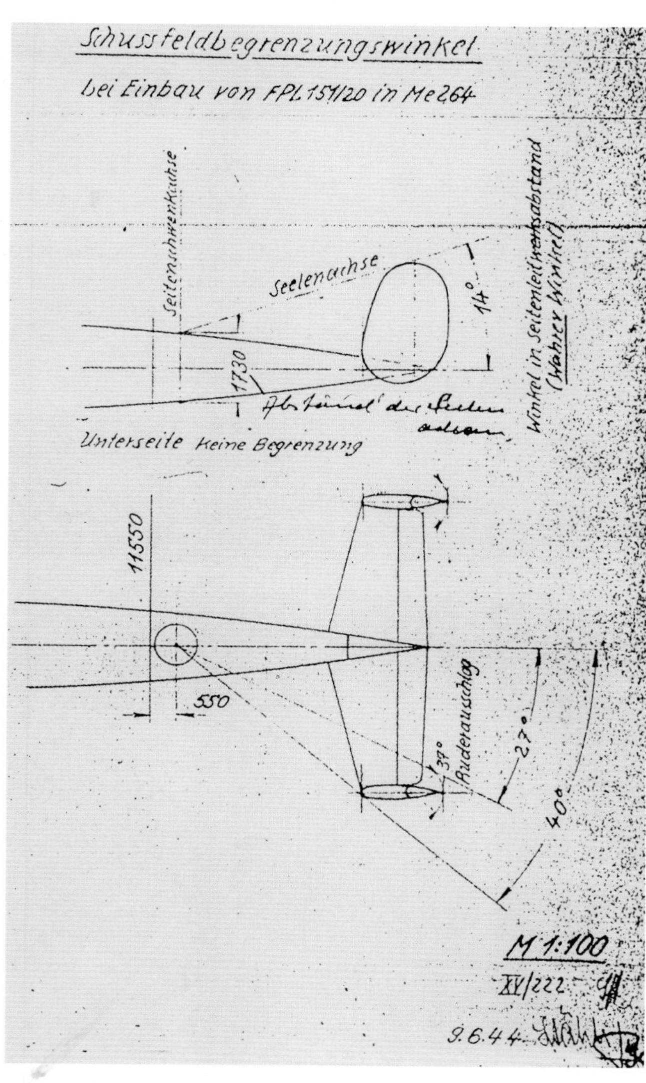

A Messerschmitt sketch dated 9 June 1944 showing the field of fire of the upper rear FPL 151/20 gun turret.

For his part, *Oberst* Petersen remained resolute as to his opinion of the Me 264. On 18 July 1944, he issued a blunt report from Rechlin:

> '*The aircraft in its whole conception is not useful on the grounds of its too-high wing loading, the complicated nature of its undercarriage, the exceptional length of runway required for take-off and the need for rocket boosters to assist in the endeavour, and the poor defensive armament. The basic idea of developing and then building the machine in a purpose-built fighter factory lacking any heavy aircraft experience is flawed and in our opinion it cannot be produced today without prejudice to the urgent Me 262 fighter production at Messerschmitt. The promised dates seem impossible to keep as still no kind of test data is available.*'[41]

However, later that very morning, it was the Americans who delivered the decisive blow to the Me 264 programme, when more than 500 B-17s and B-24s of the US Fifteenth Air Force, operating in bad weather from their Italian bases, attempted to strike at the aircraft production facility at Manzell and at Memmingen airfield. A week earlier the heavily armed and armoured Fw 190 interceptors of IV.(*Sturm*)/JG 3 had arrived at Memmingen from Illesheim under the command of *Major* Wilhelm Moritz and 45 of its fighters were scrambled to intercept the raid. Assembling over Holzkirchen, the *Gruppe* headed south towards the American bombers which were reported to be approaching from Innsbrück and Garmisch. However, the adverse weather prevented the greater part of the Memmingen strike force from hitting the target, forcing two bomb groups and their escort to turn back and another to bomb an alternative target. But the Fortresses of the 483rd Bomb Group became separated, failed to receive the recall message and pressed on alone without their P-38 escort. As the 483rd doggedly neared the Starnberger See at around 10.50 hours, Moritz' Focke-Wulfs struck and 14 of the Group's Fortresses were shot down by the *Sturmgruppe*. Nevertheless, those B-17s which survived the fighter assault made it to Memmingen and dropped their fragmentation and incendiary bombs across the target hitting several of the airfield's installations. Eighty per cent of the airfield buildings were hit in the devastating attack; three hangars were destroyed, and another two heavily damaged, as were the workshop, mess hall, and accommodation. One hundred and seventy personnel were killed and another 140 wounded. An estimated 35 aircraft were destroyed or damaged on the ground, including the Messerschmitt Me 264 V 1.[42]

Top 18 July 1944: smoke drifts across Memmingen airfield shortly after the devastating USAAF attack.

Above Two Luftwaffe NCOs, members of IV.(Sturm)/JG 3, survey the damage inflicted to Memmingen airfield by the American bombing raid of 18 July 1944. An estimated 35 aircraft were destroyed or damaged on the ground, including the Messerschmitt Me 264 V 1. The attack signalled the end of any further meaningful development of the 'Sudeten'.

As Karl Baur made his way into the smouldering remains of the '*Sudeten's*' hangar after the raid, he realised that the extent of the damage inflicted to the airframe was too severe to be repaired and that having flown only 38 hours 22 minutes in 52 flights, the aircraft would have to be written off.[43]

For Willy Messerschmitt and his dreams of building a long-range aircraft, the loss of the only fully-assembled Me 264 prototype on 18 July 1944 must have been a blow. Yet, as with so many other military and technical projects developed in the Third Reich in 1944, despite the prevailing adverse war situation and worsening conditions on the home front, the physical destruction of the V 1 did not signal the end of the Me 264 programme.

In what seems yet another contradiction in the story of the aircraft, on 26 July the *Kommando der Erprobungsstellen* authorised the establishment of '*Sonderkommando Nebel*' ('Special Detachment Nebel') under the command of *Hauptmann* Wolfgang Nebel. The unit was set up specifically to oversee the further development of the Me 264 V 2 and V 3 prototypes and to assess the best use for them. It was comprised of engineers and personnel drawn from the staffs of the *General der Aufklärungsflieger* and the *General der Fliegerausbildung* and its first headquarters was at Offingen, the location of the V 2 and V 3 airframes.

The actual accomplishments of the *Kommando* remain unclear, but in August 1944, it is possible that it commissioned *Professor* Losel of the firm of Osermaschinen GmbH to carry out the design and development of a steam turbine power unit for an aircraft which was to be rated at 6,000 hp at 6,000 rpm with a weight to power ratio of 0.7 kg/hp and a consumption of 190 grams/hp/hour. An Me 264 airframe – possibly the V 2 – was to have been placed at the disposal of the firm, but it was apparently destroyed in an air raid before experiments got under way.

Two forms of propeller were envisaged; one at 5.3 m diameter and revolving at 400-500 rpm and the other at 1.98 m diameter revolving at 6,000 rpm.

The system consisted of four capillary tube boilers, each one metre in diameter and 1.2 m high; a boiler feed water pump and auxiliary turbine; a main turbine, 0.6 m in diameter and 1.82 m in length; a combustion air draught fan, condenser, controls and auxiliaries. By the end of the war, many of these components had been manufactured and were ready for use including one of the boilers, the turbine blades and auxiliaries, such as the combustion air draught fan and condenser pump. Work had also started on the auxiliary and main turbine.

The first system was designed to use 65 per cent solid fuel (pulverised coal) and 35 per cent liquid fuel (petrol), but it was intended to use liquid fuel only when it became available in quantity. Osermaschinen claimed that the advantages of steam turbine power were:

Constant power at varying heights
Capacity for 100 per cent overloading, even for long periods
Full steam output attained in 5-10 seconds
Not sensitive to low temperatures
Long life and simple servicing requirements
Simple and quick control

The system lent itself to incorporation within an airframe, since it could be broken down into separate components.[44]

Other initiatives in which *Sonderkommando Nebel* may have been involved included further investigations into de-icing systems for the long-range reconnaissance version of the Me 264; enhancements were foreseen for the wing, tail fins, the induction air scoops on the engines, airscrews, pressure heads and cockpit glazing. To this end, data taken from earlier experiments with the Me 210 were used. It was calculated that, in total, 36.74 m of 20 mm wide continuously heated parting strips would be needed for the aircraft.[45]

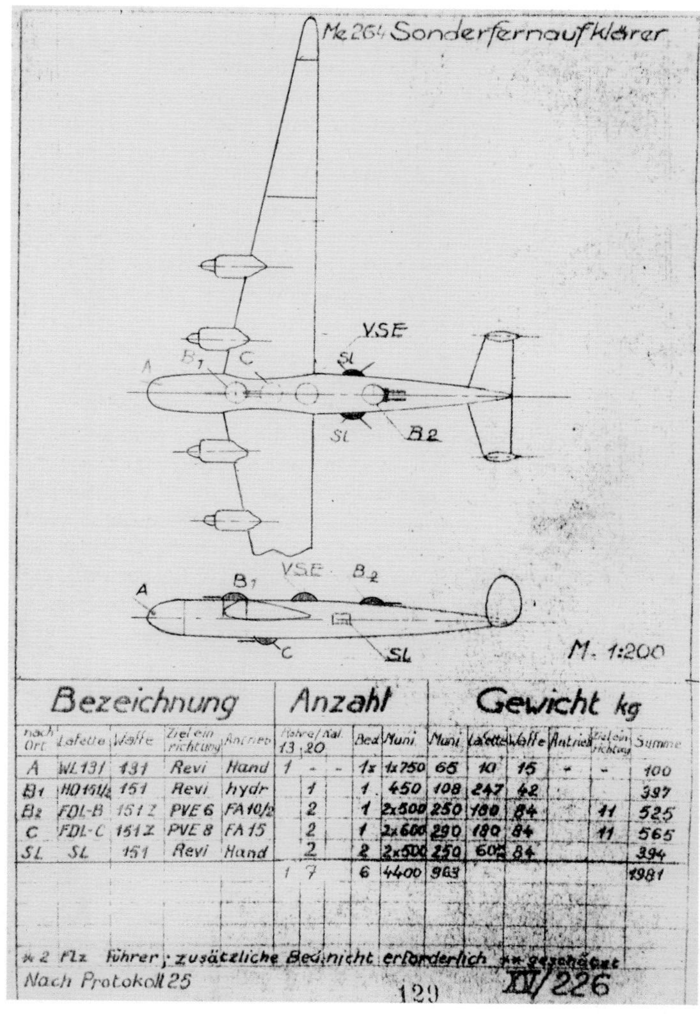

Me 264 Sonderfernaufklärer

M. 1:200

Bezeichnung					Anzahl					Gewicht kg					
nach Ort	Lafette	Waffe	Zielein richtung	Antrieb	Rohre/Kal. 13	20		Bed	Muni	Muni	Lafette	Waffe	Antrieb	Zielein richtung	Summe
A	WL131	131	Revi	Hand	1	-	-	1x	1x750	65	10	15	-	-	100
B1	HD151/2	151	Revi	hydr		1		1	450	108	247	42			397
B2	FDL-B	151 Z	PVE 6	FA 10/2		2		1	2x500	250	180	84		11	525
C	FDL-C	151 Z	PVE 8	FA 15		2		1	2x600	290	180	84		11	565
SL	SL	151	Revi	Hand		2		2	2x500	250	60	84			394
					1	7		6	4400	963					1981

* 2 Flz. Führer; zusätzliche Bed. nicht erforderlich ** geschätzt

Nach Protokoll 25

129 XV/226

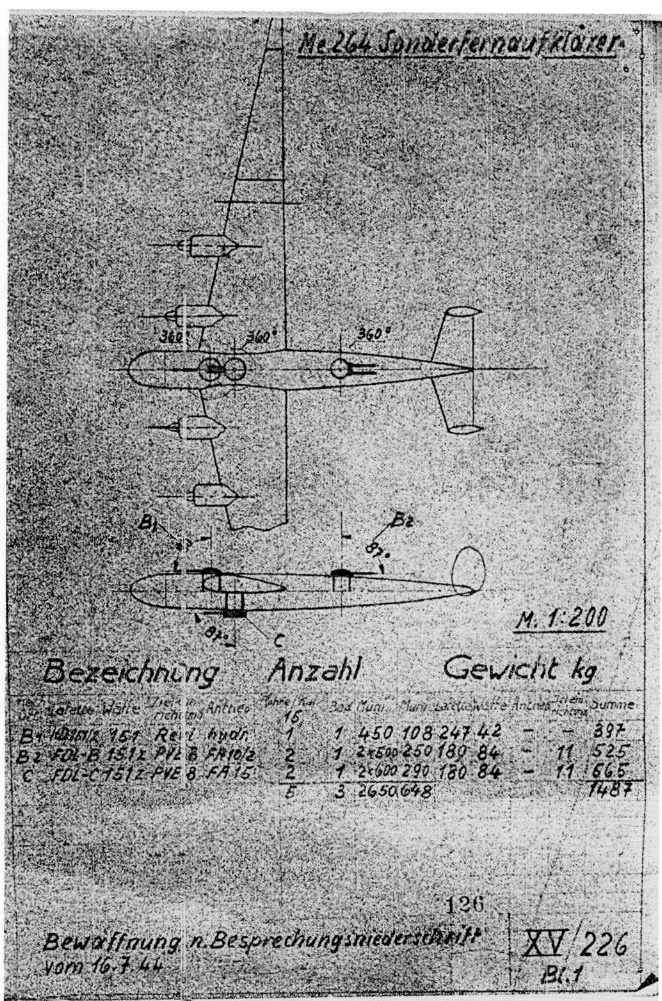

Me 264 Sonderfernaufklärer

M. 1:200

Bezeichnung					Anzahl			Gewicht kg					
nach Ort	Lafette	Waffe	Zielein richtung	Antrieb	Rohre/Kal. 15	Bed	Muni	Muni	Lafette	Waffe	Antrieb	Zielein richtung	Summe
B1	HD151/2	151	Revi	hydr.	1	1	450	108	247	42	-	-	397
B2	FDL-B	151 Z	PVE 6	FA 10/2	2	1	2x500	250	180	84	-	11	525
C	FDL-C	151 Z	PVE 8	FA 15	2	1	2x600	290	180	84	-	11	565
					5	3	2650	648					1487

126

Bewaffnung n. Besprechungsniederschrift vom 16.7.44

XV/226 Bl. 1

Above and above right Two diagrams produced in July and August 1944 showing weapons positions and armament types, calibres, sights, quantities, number of rounds and weights for the long-span Me 264 Sonderfernaufklärer.

Below All that was left of one of the remaining prototypes of the Me 264 – either the V 2 or V 3. This bare metal fuselage seems to have slipped from its tethered storage in what appears to be a timber building. Internally at least, construction seems not to have been at an advanced stage by the time this photograph was taken and it is possible the location is Offingen, where the second and third prototypes were under the care of Sonderkommando Nebel.

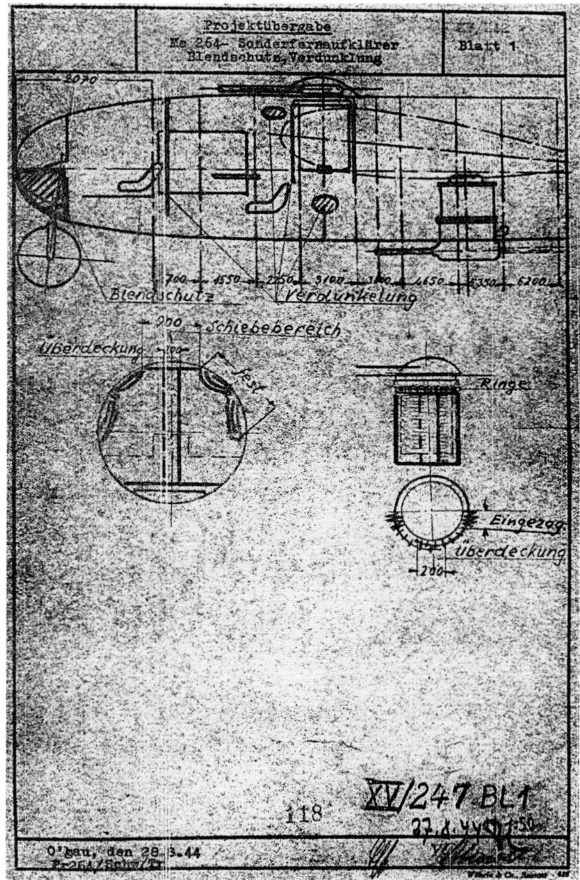

Projektübergabe
Me 264- Sonderfernaufklärer
Blendschutz, Verdunklung

Blatt 1

O'gau, den 28.8.44
Pr264/Schw/Tr

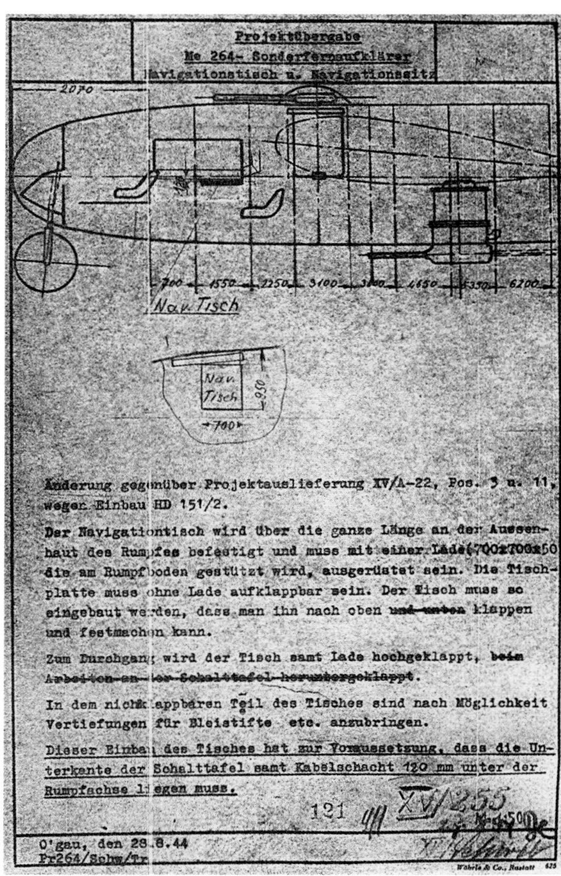

Projektübergabe
Me 264- Sonderfernaufklärer
Navigationstisch u. Navigationssitz

Änderung gegenüber Projektauslieferung XV/A-22, Pos. 3 u. 11, wegen Einbau HD 151/2.

Der Navigationtisch wird über die ganze Länge an der Aussenhaut des Rumpfes befestigt und muss mit einer Lade (700x700x50), die am Rumpfboden gestützt wird, ausgerüstet sein. Die Tischplatte muss ohne Lade aufklappbar sein. Der Tisch muss so eingebaut werden, dass man ihn nach oben ~~und unten~~ klappen und festmachen kann.

Zum Durchgang wird der Tisch samt Lade hochgeklappt, ~~beim Arbeiten an der Schalttafel heruntergeklappt~~.

In dem nichtklappbaren Teil des Tisches sind nach Möglichkeit Vertiefungen für Bleistifte etc. anzubringen.

Dieser Einbau des Tisches hat zur Voraussetzung, dass die Unterkante der Schalttafel samt Kabelschacht 120 mm unter der Rumpfachse liegen muss.

121 XV/255

O'gau, den 28.8.44
Pr264/Schw/Tr

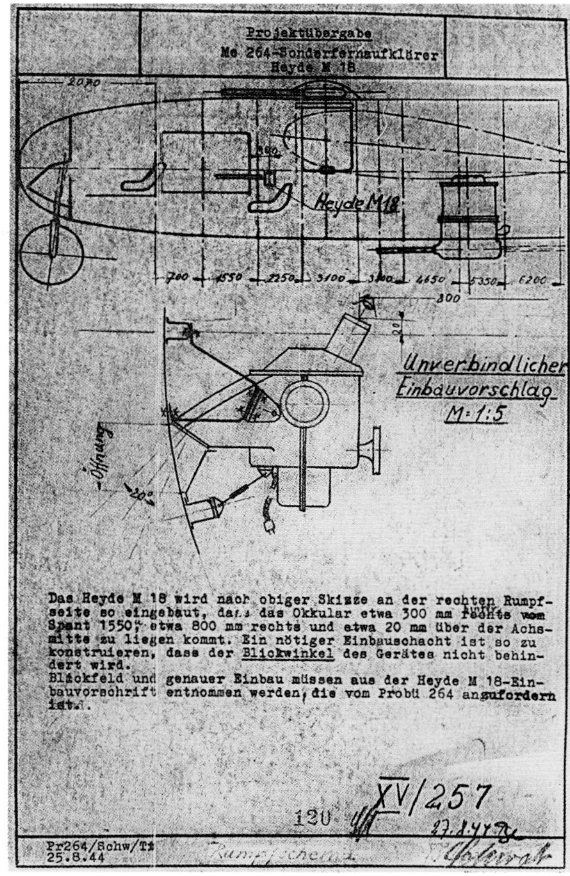

Projektübergabe
Me 264-Sonderfernaufklärer
Heyde M 18

Das Heyde M 18 wird nach obiger Skizze an der rechten Rumpfseite so eingebaut, dass das Okkular etwa 300 mm rechts vom Spant 1550, etwa 800 mm rechts und etwa 20 mm über der Achsmitte zu liegen kommt. Ein nötiger Einbauschacht ist so zu konstruieren, dass der Blickwinkel des Gerätes nicht behindert wird.
Blickfeld und genauer Einbau müssen aus der Heyde M 18-Einbauvorschrift entnommen werden, die vom Probü 264 anzufordern ist.

120 XV/257

Pr264/Schw/Tr
25.8.44

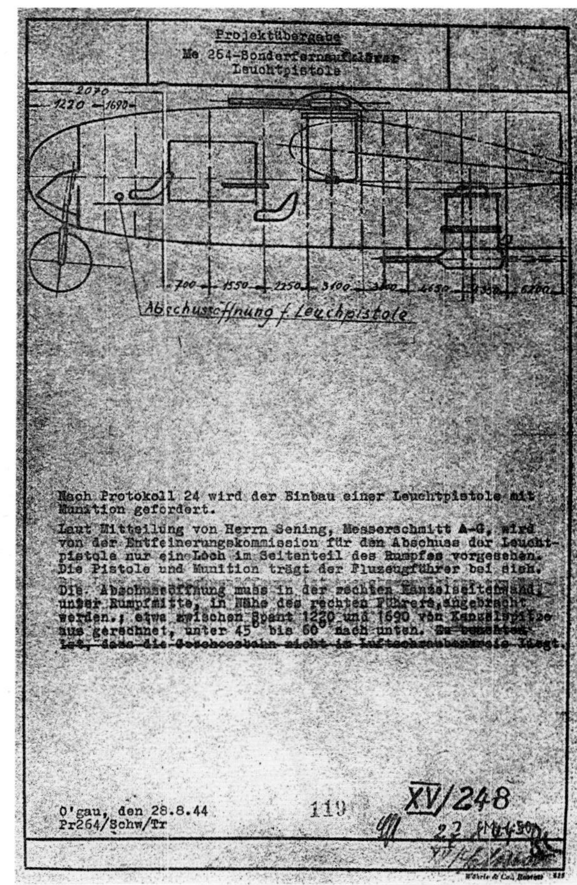

Projektübergabe
Me 264-Sonderfernaufklärer
Leuchtpistole

Nach Protokoll 24 wird der Einbau einer Leuchtpistole mit Munition gefordert.

Laut Mitteilung von Herrn Sening, Messerschmitt A-G, wird von der Entfeinerungskommission für den Abschuss der Leuchtpistole nur ein Loch im Seitenteil des Rumpfes vorgesehen. Die Pistole und Munition trägt der Fluzeugführer bei sich.

Die Abschussöffnung muss in der rechten Kanzelseitenwand, unter Rumpfmitte, in Nähe des rechten Führers angebracht werden.; etwa zwischen Spant 1220 und 1690 von Kanzelspitze aus gerechnet, unter 45° bis 60° nach unten. ~~Zu beachten ist, dass die Geschossbahn nicht im Luftschraubenkreis liegt.~~

119 XV/248

O'gau, den 28.8.44
Pr264/Schw/Tr

Four Messerschmitt diagrams dated August 1944 showing various internal forward section layouts in the Sonderfernaufklärer – from top left: anti-glare blackout curtains; top right: the position of the 950 mm x 700 mm navigation table; bottom left: positioning for the Heyde M 18 viewfinder; and bottom right: the firing port for signal flares.

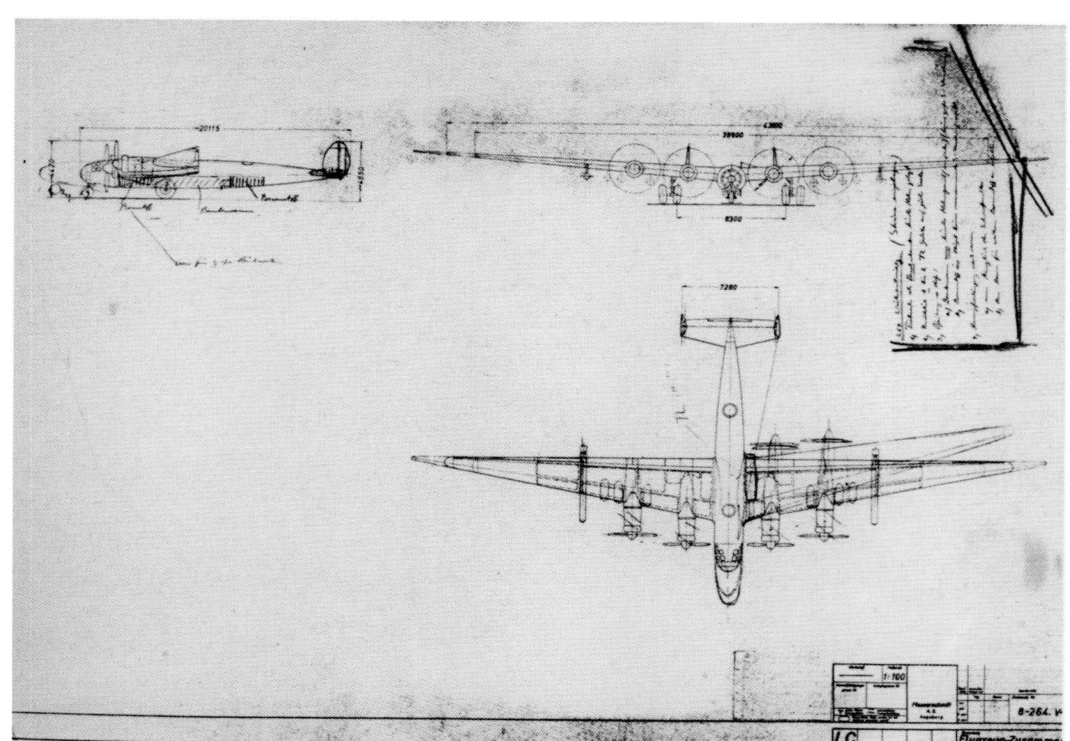

Left The drawing of the Me 264 V 1 produced by Messerschmitt A.G. seen in Chapter Three over which, in the spring of 1944, Willy Messerschmitt sketched in his ideas for an improved design featuring the installation of one of two jet engines and two of four engines with pusher propellers mounted on a swept back wing.

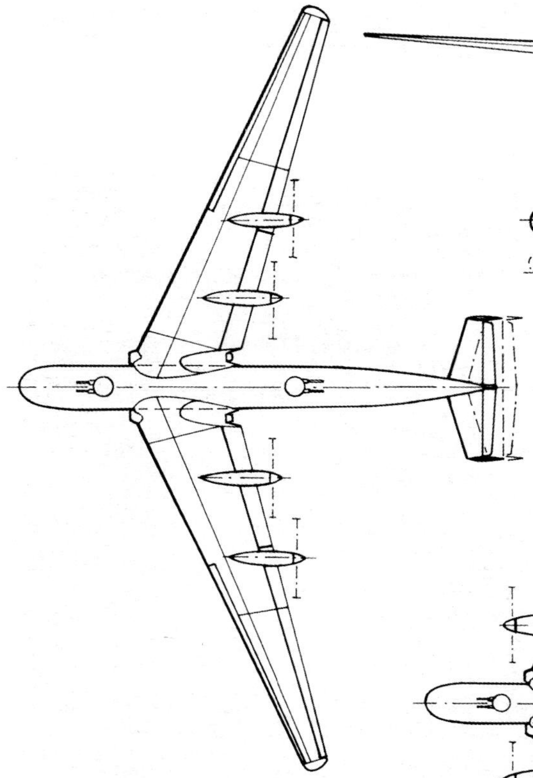

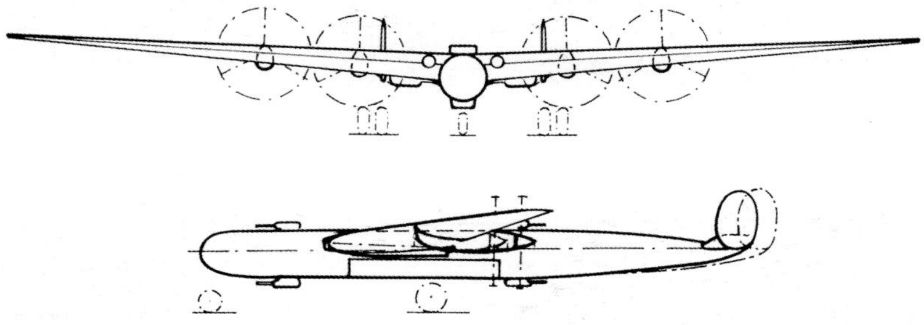

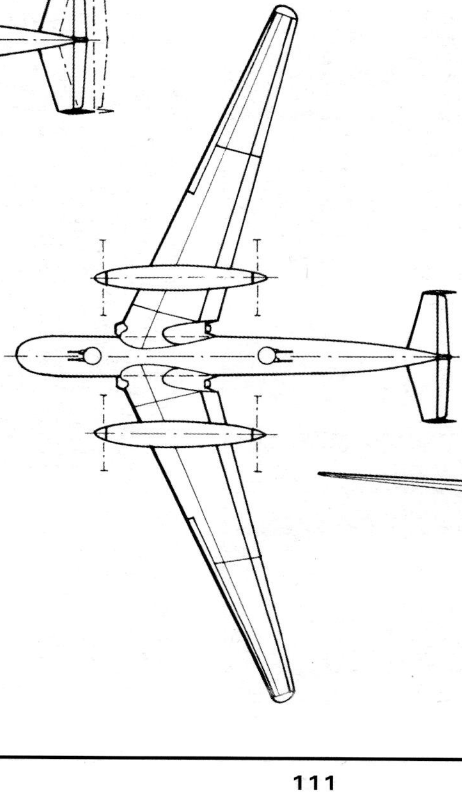

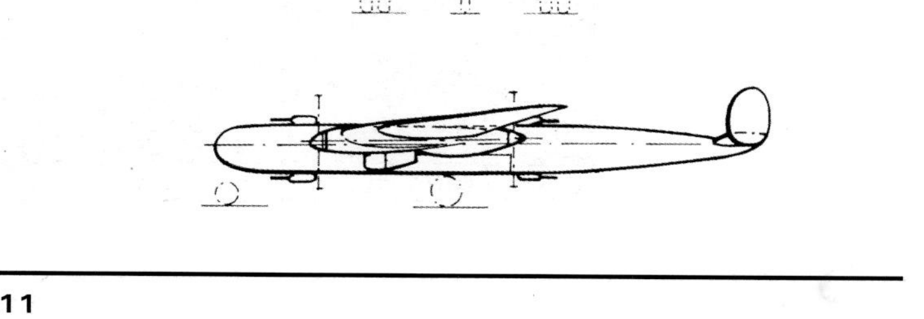

Two project drawings from late July 1944 based on Willy Messerschmitt's preceding sketch showing more precisely a configuration utilising the latest engine technology. Both projects retain the jet engines mounted in the wing root and possess a wing span of 39 metres and a length of 25 metres.

Artist's impression of Messerschmitt Me 264 swept wing project – Long-range bomber

Based on Projektbüro proposals dated July 1944 and incorporating four in-line piston engines (possibly Daimler-Benz) and two turbo-jets (possibly Heinkel HeS 011), seen in the markings of Kampfgeschwader 40.

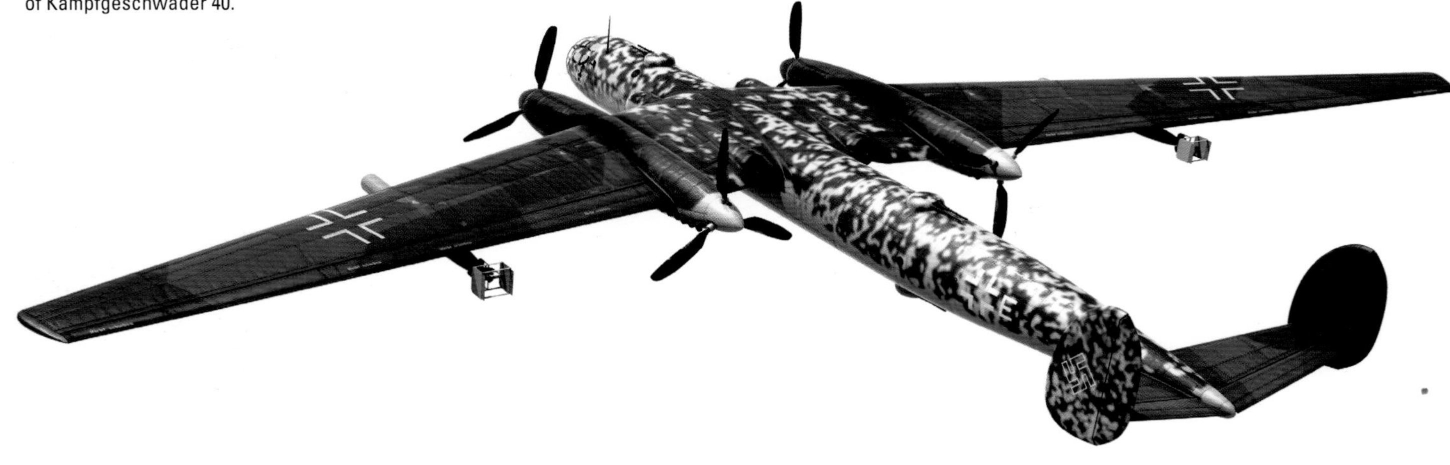

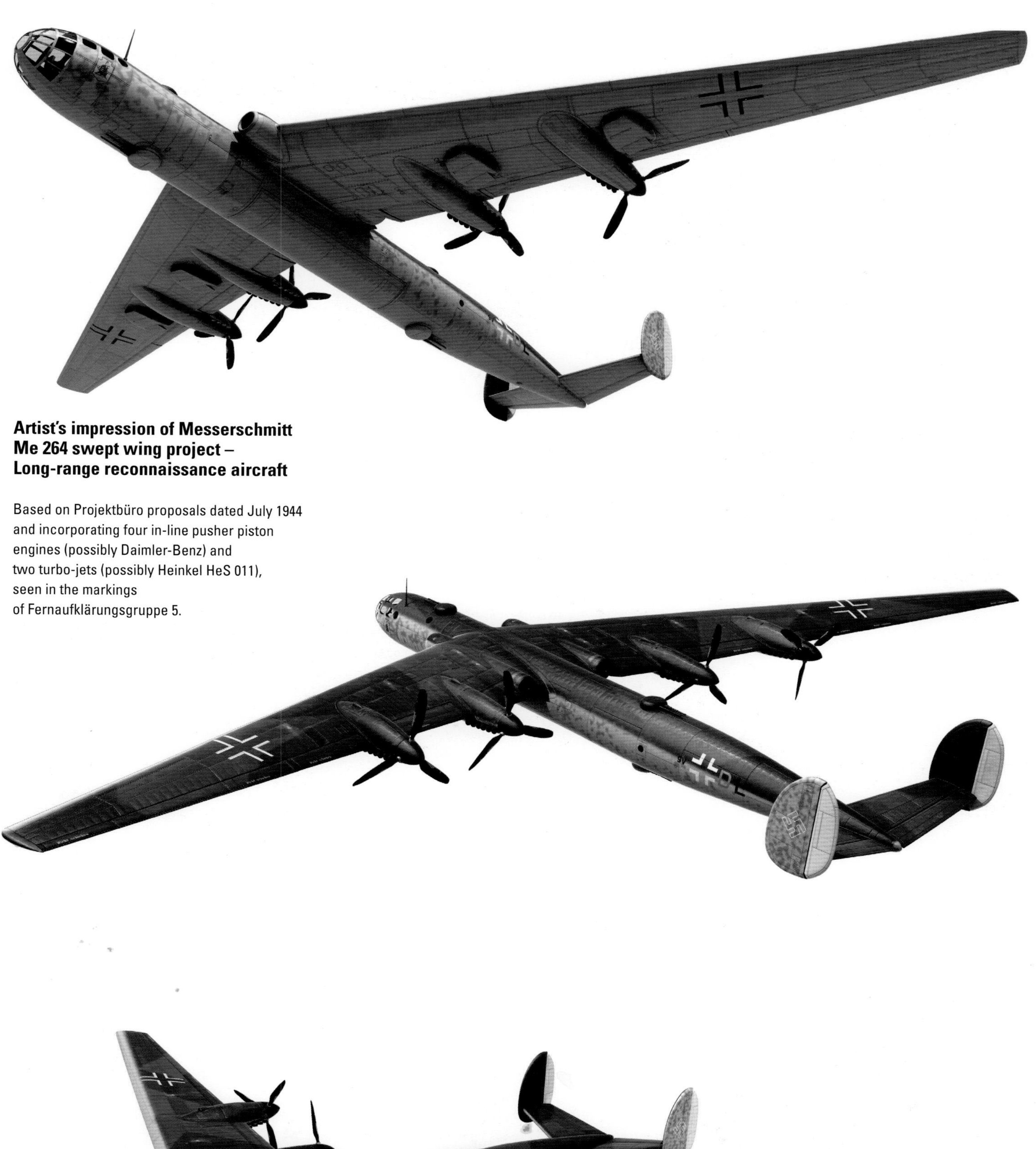

Artist's impression of Messerschmitt Me 264 swept wing project – Long-range reconnaissance aircraft

Based on Projektbüro proposals dated July 1944 and incorporating four in-line pusher piston engines (possibly Daimler-Benz) and two turbo-jets (possibly Heinkel HeS 011), seen in the markings of Fernaufklärungsgruppe 5.

On 15 August 1944, *Oberst iG* Artur Eschenauer, the *Chef der 6. Abteilung* of the OKL *Generalquartiermeister*, issued a key report on the future of both the Me 264 and the activities of *Sonderkommando Nebel* addressed to the Technical Officer on Göring's personal staff, the OKL *Führungsstab* and the *General der Aufklärungsflieger*, *Generalmajor* von Barsewisch.

According to Eschenauer: '*The removal of the Ju 290 and Ju 390 from the long-range reconnaissance programme has left a hole which needs to be closed. It is possible that the large amount of assembled parts still available would be enough to build 20-30 Me 264s and make them operational. In order to actually realise this, personnel from the* General der Aufklärungsflieger *and* General der Fliegerausbildung *should be placed with* Sonderkommando Nebel *which will have the task of coordinating the testing and deployment of the Me 264.*'

In a somewhat contradictory manner, Eschenauer also went on to state that no 'large' completed parts still existed with which to start construction, and that 80 per cent of the required materials had been destroyed. He also foresaw that the Me 264V 2 would not be ready until February 1945 and that this aircraft, weighing 50 tons and powered by four BMW 801 engines, would be armed with one MG 131 J, and three MG 151 Zs in 'waist' positions. However, Eschenauer added that: '*The question of overall weight is not fully clear for the existing undercarriage. At the moment, the proposal is to add a droppable undercarriage, but still this has to be developed.*'

The workforce available at Messerschmitt to build the Me 264 was noted at 80 'constructors' of which half were foreign with a further 20 brought in from Heinkel. If the *Technisches Amt's* plan to use more than one manufacturer was to be realised, Eschenauer opined that: '*...in order to push the matter through, a military commander should be placed in charge. For this purpose, any personnel should come from the* General der Aufklärungsflieger *or the* General der Fliegerausbildung. *One thing that would be impossible, would be to use war service soldiers in industry without the* Führer's *permission.*

'*In hindsight, because of the need to solve the technical problems and the emergency requirements, testing cannot be completed before the summer of 1945, just to have one Me 264 available. At this point in time, with the proposed performance, and without the latest technical modifications, this would probably not be of any interest to the* General der Aufklärungsflieger.

After the previously mentioned items and problems, Generalquartiermeister 6. Abt. *proposes the following:*

a) The complete development of the Me 264 in its present form to be started immediately
b) Due to changes to the existing aircraft production programmes, some spare construction capacity is envisaged and enough capacity is available for the development of a new long-range aircraft on the basis of the Me 264's performance, speed, and defensive weaponry which could be made available in the year 1945/46
c) The foreseen establishment of the Sonderkommando Nebel *will mean that it will need highly skilled technical personnel to complete the programme.*'

A detachment of crews from 2./*Fernaufklärungsgruppe* 5, the long-range maritime reconnaissance unit which had been operating the Ju 290 from Mont-de-Marsan in western France, was assigned to Nebel and arrived at Offingen under the command of *Hauptmann* Georg Eckl in late August. No sooner had Nebel's unit been established however, than it was given fresh orders by OKL. The unit was now to concentrate on the 'technical problems associated with new aircraft' – this meant not just the Me 264, but other new types such as the Do 335 and the planned Do 635.

Even as late as August 1944, Adolf Hitler still harboured thoughts of conducting air operations against the United States. On 5 August, he announced that he wanted to see the 'fastest possible production' of further Me 264s.[46] The idea lingered in his mind for at least another fortnight. On the 21st, during the *Führer's* morning situation conference, the subject of a long-range bombing mission against New York cropped up once again. *Generalleutnant* Werner Kreipe, the hapless new *Luftwaffe* Chief of Staff, recorded in his diary: '*Morning conference. Short discussion over the operation against New York with long-range bomber. The Navy cannot supply a U-boat for supply* [of fuel] *and collection* [pick-up of crew]. *Operation off.*'[47]

Kreipe subsequently discussed the 'New York operation' with *Admiral* Kurt Fricke of the OKM until late that afternoon and then spent further time on the telephone that evening with *Admiral* Wilhelm Meisel, the Chief of the SKL about the '*operation against the USA.*'[48]

Whether or not it was because of the lack of a U-boat, Hitler, who had more pressing war matters to deal with on the Western and Eastern Fronts, eventually forgot about the scheme and the sense of urgency petered out.

The urgency had not petered out from the direction of *Reichsminister* Albert Speer's Armaments Staff, which, on 30 August, ordered Nebel to arrange to '...build the Me 264 with utmost precipitation and by utilising all readily active industrial forces.' This, it seems, was an indirect reference to the Henschel Flugzeugwerke A.G. at Schönefeld which was brought in to build the wing centre sections for the Me 264 for delivery by late November 1944. By the end of October however, the terms of the order had changed seven times, and the Henschel firm, probably exasperated, ceased its involvement at the end of that month.[49]

In September 1944, the Me 264 became the centrepiece for a somewhat intriguing study by the famous long-distance German aviator, Hans Bertram, who had spent six years in China as an advisor to the Chinese Naval Air Service. In 1933, Bertram had set off on a long-distance flight to China over Turkey, Iraq, India and Ceylon, but his aircraft, a Junkers F 13ge/*See*, had become a total loss as a result of a monsoon whilst in the harbour at Vicagapatam. In 1932 Bertram had embarked upon a historic and extraordinary round-the-world flight in a Junkers W 33D seaplane. Starting his epic journey from Cologne in February, together with three companions, he had flown across Europe and the Middle East, over India and Ceylon, to Burma and Malaya, into Indonesia and Timor. Bertram and his crew had intended to visit German communities along the route and to find potential markets for Germany's aviation industry. Other than a forced landing at Porto Fossone, in Italy, due to blinding snow and engine trouble, the flight proceeded as planned until they reached Syria. Over the desert they had hit a sand storm, forcing them to land. In India one of the crew left the aircraft in order to reduce weight. In Batavia, Indonesia, another member of the crew left to cut back on weight and Bertram and just one other crew member then flew on, overhauling the engine in Soeeabaja, Indonesia. They then refuelled in Kupang, heading for Port Darwin on the north-west coast of Australia. However, the little Junkers soon flew into a violent storm and found it impossible to determine wind direction. Finally, with fuel low, they were forced to land on the remote coastline of north west Australia with little water and no food. Since there was no official record of their flight, initially no search party was despatched. Finally, on 23 June 1932, after 53 days, they were rescued and taken to Perth to recuperate. Bertram had the Junkers repaired, removed the floats and for the next four and a half months flew around Australia visiting several cities. Later, in 1933, he flew from Surabaya to Berlin-Tempelhof – a distance of 14,000 km – in six and a half days, returning to a tumultuous welcome. His round-the-world flight, however, would never be achieved.

Bertram's plan for the Me 264, which he called '*Projekt Ostasienflug*' ('East Asia Flight Project'), may have been inspired in some way from contacts he may have had within the RLM or others close to the Me 264 programme, who kept him informed as to the Messerschmitt's capabilities. His proposal is dated shortly after *Oberst iG* Eschenauer's report of August, in which it was projected that the Me 264 would become available in 1945. Bertram proposed a plan whereby regular courier flights using newly produced Me 264s between Berlin and Germany's allies in Tokyo, were planned for February 1945 onwards with the first test flight taking place, at the latest, in January.[50] As mentioned earlier in this chapter, there was still much dialogue between German and Japanese diplomats and senior military officials at this stage of the war, and the Japanese had shown interest in the design of the Me 264 as well.

Bertram based his plans on an Me 264 with a range of 11,500 km powered by four BMW 801 E or G engines with six gun positions and additional armour in the cockpit area. The crew would comprise ten: a '*Kommandant*', two pilots, a navigator, two radio operators, two engineers and two gunners. The aircraft would require four rocket-assisted take-off units, and a runway length of 1,800 metres. Weight was calculated as follows:

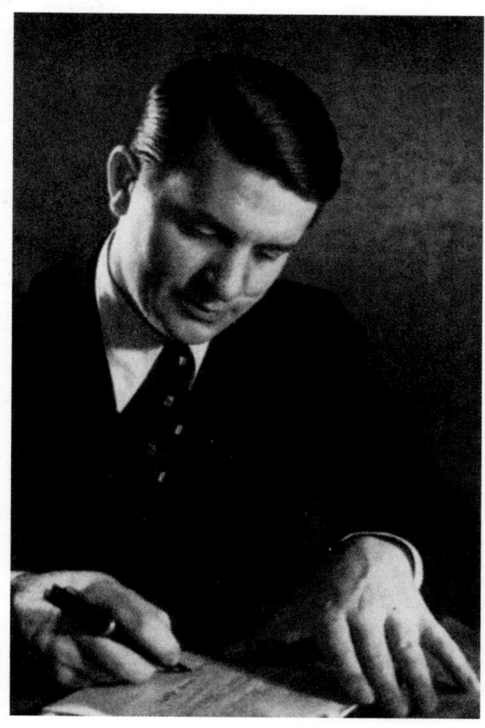

In September 1944, the famous long distance German aviator, Hans Bertram, wrote a detailed report entitled 'Projekt Ostasienflug' in which he proposed using the Me 264 on courier flights between Germany and Japan, ostensibly with the purpose of transporting 'passengers, luggage and freight'.

Outward flight (following jettisoning of RATO units):

Aircraft, equipment, crew and spare parts:	25,000 kg
Fuel:	20,250 kg
Load (passengers/luggage and freight):	5,000 kg
Additional payload:	2,750 kg
	53,000 kg

The aircraft would climb at a rate of one metre per second at 315 km/h at 2,000 m and 400 km/h at 6,000 m. Average cruising speed was calculated at 350 km/h with a flight duration of 33 hours. For the return flight, there would be a reduction in fuel of 750 kg, plus the absence of the outward load, so that the start weight would be nearer to 46,250 kg. The required runway length would be 1,350 m.

Bertram proposed two non-stop routes to Tokyo. The first, slightly longer at 9,200 km, but within the Me 264's range, was from Berlin, across Austria, Hungary, Rumania, the Black Sea, Turkey, Iran and Afghanistan, then over the Pamir mountain range, heading east for Sinkiang in China and Inner Mongolia. This would leave a range margin of 2,300 km. Having started from Berlin, the aircraft would make an interim stop at Linz-Hörsching in Austria, which, in 1945, was known to have had a 1,700 m x 50 m concrete runway – not quite long enough to equate to Bertram's calculations. Bertram's schedule was based on the aircraft taking off from Hörsching at sunset, to fly over Hungary and Rumania at night – avoiding large towns – and across the Black Sea, Turkey and Iran. By sunrise, the Me 264 would be somewhere between Tehran and Kabul. The second night would be spent overflying war-torn, Japanese-occupied China, landing in the morning at Peking to refuel after 26.5 hours in the air.

The second route, at 8,100 km, was as per the first, but from Afghanistan the course would run south-east over the Khyber and along the southern Himalayas in India, on to Rangoon in Burma. Allowing for a safety margin of an extra 2,025 km, the total distance would be 10,125 km. Using this route, the aircraft would adopt the same flight path from Hörsching to Afghanistan, but in the early daylight hours of the second day, the route would veer south over the Khyber Pass. The second night would be spent in flight over India and Burma where there was also heavy fighting between the British and Japanese.

However at the end of the first night, the crew had the option of flying to either Peking or Rangoon. There was also the option of making intermediate landings at either Bauto, 700 km west of Peking, which would reduce flying time by two hours, or at Akyab, 500 km north-west of Rangoon, representing a one and a half hour reduction in flying time.

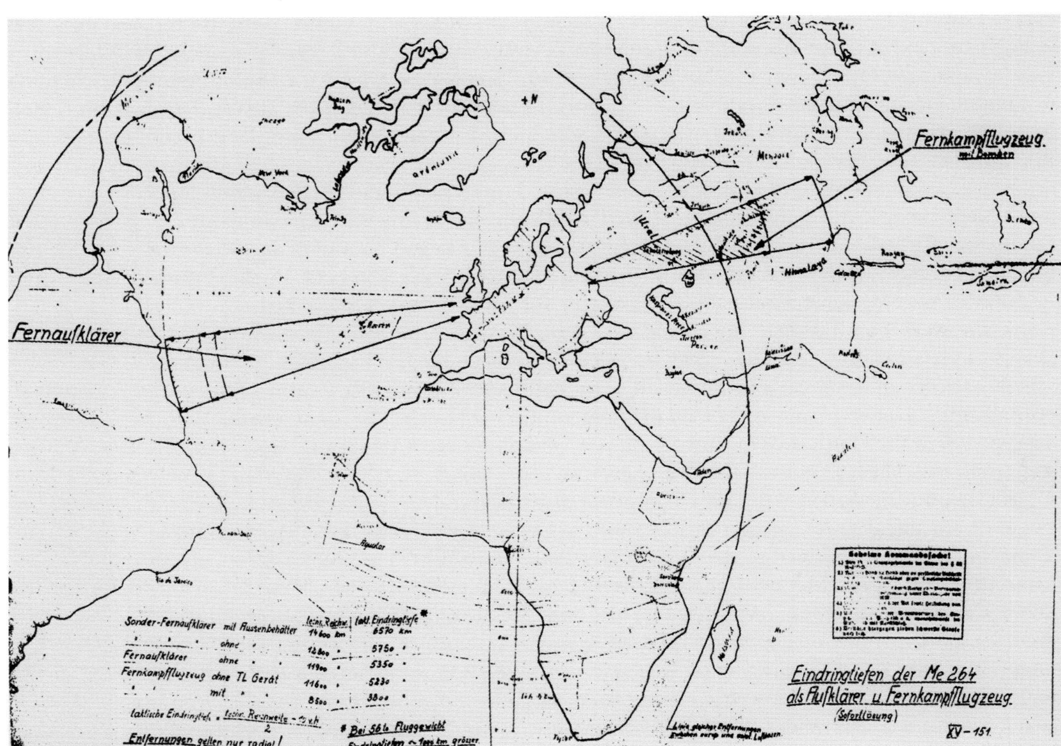

An undated map showing the planned operational radii of the Me 264 reconnaissance versions (to the west) and long-range bomber versions (to the east). Ranges of 14,600 km with drop tanks are given for the Sonderfernaufklärer and 8,500 km for the Fernkampfflugzeug fitted with additional jet engines.The Sonderfernaufklärer could reach the United States East Coast while the Fernkampfflugzeug was seen as reaching as far east as the Himalayas.

As far as is known, nothing more came of these plans, but why such calculations were made at this stage of the war to fly 'passengers, luggage and freight' from Berlin to Tokyo is open to conjecture! Interestingly, in a post-war British War Department report on German technical aid to Japan dated 31 August 1945, it is revealed that on 15 September 1944 '*... the Japanese military attaché to Berlin was able to inform Tokyo that as a result of a conference held with the Messerschmitt company, he was ready to come to an agreement for the services of three Messerschmitt technicians –* Dipl.-Ing *Rolf von Chlingensperg,* Dipl.-Ing *Riclef Schomerus and August Bringewald, the* Typenbegleiter *for the Me 262. The first two were to direct designing of short-range fighters and long-range bombers, while Bringewald would have the chief task of directing manufacture of the Me 262. In addition the technicians were to carry with them from Germany preparatory materials and data and were to be employed in Japan for a period of two years. Among those materials were to be important design data on short-range fighters and long-range bombers such as the Me 209, Me 309 and Me 264 – the direction of designing of which presumably was to be the responsibility of Chlingensperg and Schomerus... For their duties in the Far East, the technicians were to carry with them appropriate documents. Among the documents specified in the contract were those covering the Me 209, Me 309, Me 264, Me 262, Me 410 and Me 323...*'[51]

By early September 1944, *Generalmajor* von Barsewisch and *Admiral* Dönitz were forced to accept that with the imminent loss of vital airfields in France and the increasing superiority of the Allied air forces, the ability to conduct any form of cooperation between the Me 264 and the U-boat fleet was virtually impossible. On the 7th, a last-ditch meeting was held in Berlin to discuss in what way – if at all – anything could be done with the '*Sudeten*' project on a practical level. In attendance from the Navy were Dönitz and members of the Staff of the BdU, while from the *Luftwaffe*, there was von Barsewisch and *Hauptmann* Müller from the staff of the *General der Aufklärungsflieger*, together with *Major* Hermann Fischer, the commander of *Fernaufklärungsgruppe* 5, who had been asked to attend because of his recent practical experience in flying long-range maritime reconnaissance operations over the Atlantic with the Ju 290. It is quite possible that *Hauptmann* Nebel was present as well, but this is not certain.

The general attitude towards the Me 264 was lukewarm. There was little enthusiasm over an aircraft which would be expensive to produce and be available only in very small numbers by the end of the year and which, with its relatively weak defensive armament, would be hard-pressed to defend itself against the increased numbers of Allied day and night fighters operating over France, Germany and the Bay of Biscay.

That day, the war diary of the SKL noted: '*The Me 264 is too slow and therefore for long-distance maritime reconnaissance, unusable. It should be replaced by the Do 335, capable of reaching a maximum speed of 725 km/h at an altitude of 8,000 m and to cover a total range of 9,600 km.*'[52]

This seems in stark contrast to the demands of Saur, Eschenauer and Speer, and yet in the view of von Barsewisch it was only the proposed twin-fuselage Do 335 Z (*Zwilling*) which would be able to assist the Navy when it planned to resume a meaningful level of U-boat warfare in the late spring of 1945. The big Dornier, formed from the idea of mating two standard Do 335 fuselages to create more fuel capacity in a new wing centre section, was expected to possess a range of nearly 7,000 km, enough at least to reach as far as the north of Ireland and the St Georges Channel.

Recent suggestions from Willy Messerschmitt in which he proposed enhancing the Me 264's performance and extending its range by using jet engines were dismissed by von Barsewisch. Jet engines would simply consume much more fuel and thereby decrease range so dramatically that the aircraft would struggle to make it more than 2,500 km into the Atlantic – totally insufficient for long-range patrols.

Yet at another conference on the 10th between officers from the staffs of von Barsewisch and Dönitz, it was concluded that limited numbers of Me 264s *could* still be operated from bases in Norway, since, with increased armament, it was felt the Messerschmitt *would* be able to operate with reasonable safety over England, France or the western Mediterranean.

This must have served to keep the flame of hope flickering for the Me 264 construction programme; on 5 September 1944 *Generalstabs-Ingenieur Dipl.-Ing.* Roluf Lucht was appointed chairman of the Emergency Aircraft Commission. Amongst his immediate tasks was to ensure the continuation of work on the Me 264, now a priority by urgent order of the *Führer*. On the 18th, Speer once more lent his weight to the Me 264 programme by announcing that construction would begin shortly.[53]

In the week that followed however, it would be *Admiral* Dönitz who diplomatically persuaded Hitler to cancel plans for the aircraft; in truth, the *Admiral's* task was made a little easier by the fact that the *Führer* was too preoccupied with other more pressing matters so that by this stage of the war, the question of transatlantic bombing and reconnaissance was no longer relevant or a priority to him. Thus on 23 September 1944, orders were issued from the *Führer's* headquarters that all further work on the Me 264 was to be cancelled. This was followed just under a month later, on 18 October, by a stark order from Göring stating: '*Production of the Me 264 is herewith cancelled.*'

Yet even this did not kill the project entirely. In December 1944, as German forces fought their last significant counter-offensive of the war in the snow-covered forests of the Ardennes and the Soviet Army approached Budapest, Messerschmitt engineers continued their work on developing designs for the Me 264 at the '*Metalbau Offingen*'. It is possible that Hans Bertram's proposals from September had been under consideration, for one idea being worked upon at this time was a courier version of the Me 264, with a range of 12,000 km and a payload of 4,000 kg. But since virtually all the necessary parts and components for such a project had by now been scrapped or commandeered for other tasks, it seems likely that the whole idea was a little more than a ruse to prevent Messerschmitt employees being caught by military conscription teams who were combing government and civil institutions for vitally needed personnel for the war fronts.

Meanwhile, on 5 December 1944, *Sonderkommando Nebel* at Offingen was tasked by *Generalmajor* Ulrich Diesing, the *Chef der Technischen Luftrüstung*, with urgently taking charge of development policy with regard to all long-range reconnaissance, long-range *Zerstörer* and long-range fighter types, particularly to assist the new U-boat campaign for 1945. The *Kommando* was to cease any special priority it gave to the Me 264.[54] This was reinforced on 16 January 1945, when the *General der Aufklärungsflieger* delineated the objectives of *Sonderkommando Nebel* as the development and operational trials of aircraft for special long-range missions for the *Seekriegsleitung* and the *Luftwaffenführungsstab* as distinct from *Fernaufklärungsgruppe* 5 which was working on reconnaissance for U-boats, operations over north-east England with the Arado Ar 234 and planned operational trials with the Arado jet and the anticipated Do 335 Z.[55]

By mid-February 1945, with the worsening war situation, there were proposals to disband *Sonderkommando Nebel* – which was known to have a strength of 388 personnel, including nine officers and 44 civilian 'helpers' or auxiliaries – and assign its members to frontline service. On 12 February, at a meeting of the heads of the various operational arms of the *Luftwaffe*, it was decided to retain the *Kommando* as a small military unit intended to fulfil tasks for industry using civilian personnel. Its fate at the end of the war is not known.[56]

In the increasingly surreal atmosphere of his headquarters, Hitler seems to have rekindled his interest in long-range 'vengeance missions' by January 1945 for he was still pressing Albert Speer to deliver him 'a high-speed heavy bomber with wide range and large bomb load.'[57]

Speer himself recalled a meeting with Hitler in February 1945, a time when Allied bombs were raining down almost daily on the Reich: '*As soon as I finished my report, Saur began trying to balance out the sombre note I had struck. He spoke of a recent consultation with Messerschmitt and drew some first sketches of a new four-motored jet bomber from his briefcase. Although building a plane with sufficient range to reach New York would have taken years even under normal conditions, Hitler and Saur went into raptures over the dire psychological effects of an air raid upon the skyscraper canyons of New York.*'

However, the reality was that Hitler's vision of attacking the 'skyscraper canyons' had been doomed for a long time. The environment of damaging rivalry, personality conflicts, fear, mistrust and lack of co-operation at senior level within certain quarters of the German military and the aviation industry that existed throughout much of the war did not help. These factors, combined with an inability to mesh the required engine technology with airframe development, as well as the devastating impact of the Allied strategic bombing offensive, served to destroy such prospects.

Perhaps it was *Oberst* Bernd von Brauchitsch, Göring's *Luftwaffe* adjutant, who summarised best the damning attitude toward the German heavy/long-range bomber/reconnaissance aircraft programme when he wrote in the summer of 1945: '*It is extremely difficult to explain the failure in this field of development. A large number of converging factors were involved... Bomber types for which provision had been made were fairly well advanced as regard the airframe, but the engines which formed the basis of these designs had either not reached the final stage of development or were not yet ready for production. The principal reasons for this situation were no doubt insufficient technical understanding on the part of the General Staff, insufficient tactical ability and inexperience on the technical side of the service, lack of imagination, inadequate co-operation and conflicting interpretations.*'[58]

Chapter Six

Myth and Reality

'The myth of a German air threat to the USA died hard. When I first visited the Pentagon and the White House in June 1943, it was still absurdly alive and one had difficulty quashing it.'

W/Cdr Asher Lee, *Göring – Air Leader*, 1972

Myth

Given Hitler's war aims prior to December 1941, and the subsequent demanding multi-front war which Germany fought thereafter, in context, for the Germans to have labelled an aircraft or series of aircraft '*Amerika Bomber*' was – not surprisingly – to invite myth both at home and beyond the Reich. The British wartime air technical intelligence specialist and later air power theorist, Wing Commander Asher Lee, wrote: '*The myth of a German air threat to the USA died hard. When I first visited the Pentagon and the White House in June 1943, it was still absurdly alive and one had difficulty quashing it.*'

As mentioned earlier in the text, the Allies first became aware conclusively of the existence of the Messerschmitt Me 264 in the spring of 1943 from aerial reconnaissance and then tried to find out more. One channel of information came through German prisoners of war. On 18 April 1944, a prisoner who had been at Lechfeld in the summer of 1943 told his British captors that he had seen an aircraft known as the 'Me 264' '*... in the open on Lechfeld airfield for several months up to August 1943 when it suddenly disappeared. It aroused the prisoner's interest owing to its reputed prodigious range; it was usually referred to as the 'USA Bomber' as it was supposed to be capable of attacking the United States, and one prisoner asserts that it had been flown to Tokyo and back.*'[1]

In mid-October 1944, British Intelligence was taken in by one prisoner who claimed that since the beginning of 1944 there had been '*...regular air travel between Germany and Japan established for the transport of high officials.*' The aircraft used were allegedly Me 264s '*... flown by old, experienced Hansa pilots.*' According to this source, 27,000 litres of fuel were needed and the course taken was supposed to have been the great circle route via Leningrad.

On 10 June 1945, a Technical Intelligence Report was issued by the United States Strategic Air Forces following 'discussions' with Germans who had worked on or been connected with the Me 264 in some way. This report claimed that the Me 264's purpose was '*... to drop propaganda in the US and also to bomb the US so that we would feel we had to keep more anti-aircraft guns in the US and not ship them to Britain.*'[2]

There is also another, more recent, myth which surrounds the Me 264 – its purported planned use as a carrier for a Nazi atomic, nuclear or 'dirty' bomb. To a certain degree, the origins of this myth may lie in research conducted by David Irving during the 1960s. Drawing from diaries and interviews conducted with former German scientists, industrialists and officials, Irving recounted a meeting which took place in Berlin in early June 1942 – the date is unclear, but it was probably sometime between the 4th and the 6th of that month.[3] The results of Irving's research were published in his ground-breaking book *The Virus House* in 1966, one of the earliest post-war histories of the Nazi atomic bomb and nuclear research programme. According to Irving, *Generalfeldmarschall* Erhard Milch was invited by *Reichsminister* Albert Speer and Karl-Otto Saur to attend an important evening conference at the headquarters of the Government-sponsored Kaiser-Wilhelm Institute for Physics (KWI) in the Berlin district of Dahlem, on the subject of scientific development – though Speer's own journal records the occasion in more layman-like vernacular as dealing with '... *atom-smashing and the development of the uranium machine and the cyclotron*.'[4] Milch accepted Speer's invitation and joined him and a glittering array of some of Germany's most powerful military men, including *Generalleutnant* Emil von Leeb, head of the Army Ordnance Office and *Admiral* Karl Witzell filling the same position with the Navy, as well as leading industrialists including *Professor* Ferdinand Porsche, the Volkswagen designer and Doctor Albert Vögler, the head of United Steel and president of the Kaiser-Wilhelm Institute.

Also attending the meeting in the high-ceilinged splendour of the Helmholtz lecture room of the Harnack building, was a group of scientists headed by *Professor* Werner Heisenberg, a brilliant young physicist, nuclear theorist and former Professor of Physics at the University of Leipzig. For the past few months, Heisenberg and his team, comprising *Professor* Otto Hahn, Director of the KWI, *Professor* Paul Harteck, a physical chemist, *Professor* Adolf Thiessen, *Dr* Kurt Diebner, an expert in uranium research for the KWI, and *Dr* Karl Wirtz, a physical chemist, had struggled with the challenge of atomic and nuclear fission.

This was an opportune moment for the scientists for it had been less than a fortnight earlier that Milch had held in-depth discussions with Friebel, von Gablenz and Schwencke of the RLM and Croneiss of Messerschmitt A.G. regarding the *'Aufgabe Amerika'*. Dietrich Schwencke had produced his extensive report on the various types of long-range aircraft available to the *Luftwaffe* and outlined the various missions with which they could be assigned, particularly with regard to targets in the United States; several of Germany's cities had by this stage suffered the effects of bombing by the RAF and the Americans had bombed Japanese cities two months earlier.

In December 1939, Heisenberg had produced a report for the Army Ordnance Office in which he outlined the prospects for fission. Heisenberg concluded that it was technically feasible to develop a controlled fission reactor and that uranium, enriched in the isotope U-235, would create a formidable new explosive '... which surpasses the power of the strongest explosive materials by several orders of magnitude.'[5]

Heisenberg examined alternative moderators with different quantities of natural uranium and predicted that graphite and 'heavy water' (water in which the hydrogen atoms possess an extra neutron in the nucleus) would be the best elements with which to slow neutrons to produce a fission in a uranium-235 nucleus rather than being captured by the more plentiful uranium-238. Heisenberg proposed a reactor of about one cubic metre filled with large quantities of the rare isotopes and up to 600 litres of heavy water, to prevent the neutrons from escaping. At this stage Germany possessed nowhere near such amounts of these items.[6] Furthermore, economic realities were stacked up against Heisenberg's research; on 5 December 1941, with the German war economy almost at breaking point, *Professor* Erich Schumann, the Director of the Military Research Office, wrote a memo to all parties involved with uranium research warning that '... *the work on the project ... is making demands which can be justified in the current recruiting and raw materials crisis only if there is a certainty of getting some benefit from it in the near future*.'[7]

Nevertheless, Heisenberg's quest for pure uranium-235 offered incredible weapons potential; as Jeremy Bernstein points out: '*If one could obtain enough nearly pure uranium-235... and compressed it into a ball, the effect of the chain reaction would be nearly instantaneous, producing an incredibly powerful explosion. In urging the Army to support isotope separation, Heisenberg pointed out that separation was the "surest method" to obtain a working reactor, and, most importantly, it was "the only method for producing explosives."*'[8]

By May 1942, German scientists had made a breakthrough by manufacturing a total of just under three and a half tons of pure uranium and from this tonnage – in powdered form – 572 kg was delivered to Heisenberg at his laboratories in Leipzig in February of that year. On 3 June, the day before the Berlin meeting, Heisenberg and another scientist, *Professor* Robert Döpel, in what was to be the fourth in a series of experiments with atomic piles, set about immersing an aluminium sphere containing the uranium, together with 140 kg of heavy water as a moderator, into a tank of water.[9] By the time the pile had been winched down into the water tank it weighed just under a ton. A radium-beryllium neutron source was then inserted into the pile's centre to measure the increase in neutrons. It was the first experiment of its kind in the world.

When Heisenberg stood up to address the audience of 50 or so people at the KWI conference in early June 1942, he chose to go straight to how an atomic 'bomb' could be used for military purposes. At the mention of the word 'bomb' there was 'an audible stir in the lecture room'.[10] This introduced a very new angle to uranium research. Heisenberg went on to explain that two nuclear explosives existed: uranium-235, and element 94, or plutonium. Immediately after he had finished his talk, Speer asked Heisenberg '...*how nuclear physics could be applied to the manufacture of atom bombs. His answer was by no means encouraging. He declared, to be sure, that the scientific solution had already been found and that theoretically nothing stood in the way of building such a bomb. But the technical prerequisites for production would take years to develop, two years at the earliest, even provided the programme was given maximum support*.'[11]

Seen here delivering a speech at the Luftwaffe Test Centre at Rechlin, Reichsminister Albert Speer, the Minister for Armaments and War Production, asked Heisenberg how nuclear physics could be applied to the manufacture of atom bombs.

Professor Werner Heisenberg, head of the wartime German scientific research team working on atomic and nuclear fission. He told Generalfeldmarschall Erhard Milch that an atom bomb would be '... about as large as a pineapple.'

During the subsequent question and answer exchange following Heisenberg's address, *Generalfeldmarschall* Milch asked the scientist just how large a nuclear bomb would need to be in order to destroy a large city. Gesturing with his hands, Heisenberg apparently replied '... about as large as a pineapple.'

Recollections vary on this point. In a post-war account, Heisenberg was quoted as saying: '*After my lecture* Generalfeldmarschall *Milch asked me approximately how large a bomb would be, of which the action was sufficient to destroy a large city. I answered at that time, that the bomb, that is essentially the active part, would have to be about the size of a pineapple.*' Another present at the meeting recalled Heisenberg using the phrase 'about as big as an *ananas*' (the German for pineapple), while Heisenberg told Irving: 'I said in an offhand way – as big as a football, or like a coconut, it would be something like that.'[12]

Milch took a deep intake of breath, but Heisenberg quickly added that the prospect of Germany developing such a bomb in the foreseeable future was an 'economic impossibility', despite the fact that the Americans could well have a uranium pile very soon and perhaps even a uranium bomb within two years.[13] This meant that there would be no immediate threat from the Americans before 1945.

Whatever the case, there is thus evidence that firstly, Heisenberg knew in June 1942 the size and critical mass of an atomic bomb and that secondly, such a bomb could have fitted easily within the bomb bay of an Me 264. However, Irving tells us that Milch left the building '... *unimpressed, and two weeks later formally authorised the mass-production of a simple, unsophisticated weapon to become notorious as V-1, the flying bomb.*'[14]

As Irving recounts: '*That the German nuclear scientists should have failed to fire Speer's imagination with the possibilities of atomic fission was their greatest shortcoming. They had their opportunity at their meeting with him in June 1942 but they wasted it.*'[15]

German atomic and nuclear research stumbled on for the rest of the war. Some writers have speculated that had Hitler been in possession of a functioning bomb by 1941 – which, as is clear from the above, would not have been the case – he would have presented the United States with an ultimatum: that America should adopt a policy of isolationism and thus leave Germany to pursue its war aims in Europe and Russia, or risk the dropping of an atomic radiation bomb on one or even more of its major cities[i]. If that had been the case, these writers argue, a city such as New York may have suffered as many as one million of its population killed, and a state of apocalyptic catastrophe would have ensued. Furthermore, if the United States had succumbed to German demands for a policy of isolationism, such a stance would have drastically impeded Britain's ability to continue the war.[16]

From the point of view of this study, what needs to be addressed is not so much if and when the German atomic programme would have been realised to the point of a functioning bomb, but how it would have been *delivered*. The simple facts are that by the time Heisenberg and his team were in a position to conduct a final criticality experiment on a uranium pile, it was late February 1945 and the German nuclear research programme had been evacuated from Berlin to southern Germany. At a cave in the Swabian village of Haigerloch, Heisenberg's scientists had gathered one and a half tons of uranium cubes, one and a half tons of heavy water, ten tons of graphite blocks and some cadmium metal in case there was any instability with the planned chain reaction. By this time however, all but the most dedicated Nazis knew that the war was reaching a conclusion which did not favour Germany. Furthermore, of the two completed aircraft capable of *delivering* an atomic bomb to New York and returning, the Messerschmitt Me 264V 1 had been destroyed by Allied bombing seven months earlier and the Junkers Ju 390V 1 was languishing at Dessau without any propellers. The latter aircraft was destroyed by the Americans in April 1945. Any plans to progress with the unfinished airframes of the Me 264V 2 and V 3 came to nothing following the order in January 1945 to *Sonderkommando Nebel* at Offingen to abandon further work on them.

The last hope remained in the 'next generation' of '*Amerika Bomber*' projects, of which the leading contender was the Horten H XVIII 'flying wing' jet bomber. Designed by Reimar Horten in December 1944, the H XVIII A was to have had a range of 11,000 km and carry a bomb load of 4,000 kg. It was to be powered by four or six Heinkel-Hirth He S 011 jet engines or eight BMW 003A or six or eight Junkers Jumo 004B turbojets. These were to be faired into the fuselage with the air intakes located in the wing's leading edge and the exhaust exiting the rear of the aircraft. However, Reimar Horten became disenchanted over the way in which the RLM handled his design, and decided to embark upon an enhanced version of the basic concept which became known as the H XVIII B. This, like the 'A', was to have been a swept back all-wing design, featuring a span of 40 metres in which a three-man crew were to have been accommodated in a bubble cockpit at the apex of the 'V'. Weighing 33 or 35 tons, propulsion was to have come from four Heinkel-Hirth He S 011 jets attached in pairs to two large underwing nacelles. A fixed multi wheeled undercarriage was mounted beneath these two nacelles and during flight the wheels were to be covered by doors. The aircraft had a range of 11,000 km and a maximum speed of 850 km, purportedly with a ceiling of 16 km and a flight duration of 27 hours. The H XVIII B would have carried a 4,000 kg offensive load to the US East Coast together with armament of two 30 mm cannon mounted in the wing leading edge, forward of the cockpit. By 23 March 1945, Horten had presented the design to the RLM.

The project received approval from Göring and considerable support from *Oberst* Knemeyer, who had

i See, for example, Geoffrey Brooks, *Hitler's Nuclear Weapons*, Leo Cooper, London, 1992

flown the twin-engine all-wing H VII at Oranienburg in November 1944 and been favourably impressed. It was likely either *Hauptdienstleiter* Karl-Otto Saur or the *Chef* der TLR, *Generalmajor* Ulrich Diesing, saw the futuristic H XVIII project as a potential carrier for an atomic or nuclear bomb, and proposed that it be built in a vast complex of bomb-proof production bunkers using predominantly slave labour at the REIMAHG (**R**eichs**ma**rschall **H**ermann **G**öring) facility at Kahla/Thüringen in the Harz Mountains. A production contract was granted on 23 March 1945 and construction was to start on 1 April 1945, but the Allied advance soon reached the area, and the project progressed no further.

When American troops captured Haigerloch on 23 April 1945 they destroyed certain items of equipment belonging to the German nuclear programme, while others were eventually shipped to the United States for further investigation. As Heisenberg wrote to *Professor* Hans Bethe: '*German physicists had no desire to make atom bombs... German research never came far enough to make a decision on the bomb.*'[17]

Reality

From the few surviving accounts, those who worked closely with the Me 264 seemed to have rated it highly. In August 1945, representatives of the British Royal Aircraft Establishment at Farnborough had a chance to interview Karl Baur in France; during the discussions, the test pilot produced some photographs: '*Baur showed us some photographs of this four-engined conventional bomber, weighing 45 tons, designed and built by Messerschmitt and first flown at the end of 1943. It seemed a very shapely machine, looking like a 'real' aeroplane as compared with the He 177. Only one prototype was built... It was flown extensively by Baur – about 70 flights – and was quite highly developed. Early in its career Baur dived it to 750 kp/h without meeting any difficulty. Remarkably few teething troubles were experienced on stability and control... In Baur's opinion "speaking impartially, and not as a Messerschmitt man," it would have proved an absolutely first class aircraft.*'[18]

On 7 September 1945, *Dipl.-Ing.* Woldemar Voigt, the former head of the Messerschmitt *Projektbüro*, described the Me 264 to the Allies as being intended for '*...submarine co-operation and for leaflet dropping in the USA*', whilst Willy Messerschmitt himself described the machine as '*... a very good and cheap aircraft.*'

Why then, did the *Reichsluftministerium* or *Luftwaffe* not pursue Willy Messerschmitt's P 1061 or Me 264 designs between 1938 and 1941? While Göring, Milch, Udet, Jeschonnek, Kesselring and others are often blamed for their blindness when it came to the concept of a 'strategic', 'long-range' or 'heavy' bomber, it should be borne in mind that for the *Luftwaffe* to have cut back on fighter and medium bomber production from 1942 onwards in favour of large and expensive heavy bombers would perhaps have been even more disastrous than the entire decision of placing all emphasis on a medium bomber force. Furthermore, Germany's attempts to build 'heavier' bombers – such as the He 177 – were plagued by mechanical and technical problems, as well as a lack of engines. This is illustrated by the fact that the *Luftwaffe* dropped only three per cent (74,172 tons) of the quantity of bomb tonnage that the Allies dropped on Germany (1,996,036 tons) between 1940 and 1945.[19] In any case, by late 1942/early 1943, *fighter* production was vital for defence against the developing Allied strategic bombing offensive which threatened the Germans' chances of building anything.

To create a heavy bomber fleet would have meant the retraining of pilots and aircrew and the modernisation, re-equipment and expansion of several airfields (even the testing conditions for the Me 264 at Augsburg proved unsuitable). By the mid-war period however, the pressure on manpower and general operating conditions made this increasingly difficult and such airfields would simply have become targets for Allied bombers.[20]

This view is given some credence in the (post-war) opinions of *Generalfeldmarschall* Albert Kesselring, who as Chief of the *Luftwaffe* General Staff from 1936-37, disallowed the development of a German four-engined bomber: '*Even if the role of the* Luftwaffe *had been viewed as a strategic one, and a well thought-out production programme devised to cover it, by 1939 there would still have been no strategic* Luftwaffe *of any significance... Even if suitable aircraft had been available* [by 1940 or 1941] – *itself hardly within the bounds of possibility – we should certainly not have had them, or trained crews to fly them, in the numbers necessary for a successful and decisive air operation. It is even questionable, to say the least, whether output could have kept pace with losses.*

'*With the prevailing shortage of raw materials, the production of strategic bombers in any adequate numbers could only have been achieved at the expense of other aircraft types. One of the lessons of the Second World War was the number of aircraft and quantity of munitions it takes to dislocate the economy of a nation.*

'*Such an objective – in the first years of the war, without the additional armaments potential of adjacent states – was for Germany unattainable.*

'*So far as I can assess the position regarding raw materials, fuel and productive potential both of aircraft and trained crews, I can only say that a strategic air force would have been created too late, and the Army would have suffered for want of direct and indirect air support.*

'*How such a strategic* Luftwaffe *would have affected the course and outcome of the war is impossible to say. The fact remains that Germany's basic error was to open hostilities when she did. Given that, any criticism of the actual role that the* Luftwaffe *fulfilled can only be theoretical.*'[21]

In a statement of considerable irony contained in a post-war letter to *General der Flieger a.D.* Paul Deichmann, Erhard Milch wrote: '*The Junkers and Dornier four-engine bombers* [Do 19 and Ju 89] *were not approved for mass production, despite the fact that the test models had proved highly promising. As a result Germany had no really adequate aircraft model for use in strategic operations; without any doubt, this is one of the reasons for the failure of the air offensive against Britain and for the* Luftwaffe's *inability to provide adequate air protection for German submarines at sea.*'[22]

Appendix 1

Performance and Weight Comparisons between the four-engined and proposed six-engined Me 264 and the projected Focke-Wulf six-engined bomber (Ta 400) – all powered by BMW 801 E engines – as produced by Seifert, Hornung and Degel, Messerschmitt A.G., Augsburg, 15 May 1943

F e r n k a m p f f l u g z e u g e – V e r g l e i c h

Me 264 – F ocke Wulf – Me 264 / 6m.

Flugzeugabmaße

	Me 264 (Normalprofil)		Focke – Wulf	Me 264 / 6m Bei Reichweitenprofil u.Normalprofil	
Anzahl der Motoren	4 x 8ol E		6 x 8ol E	6 x 8ol E	
Flügelfläche F (m^2)	127,7		17o	17o	
Spannweite b (m)	43,o		42,o	47,5	
Seitenverhältnis λ	14,5		1o,4	13,3	
Länge l (m)	ohne Heckstand 2o,9	mit Heckstand 22,5	28,2	ohne Heckstand 24,35	mit Heckstand 25,o

Augsburg,den 13.5.43
Hornung

Anlage I
Blatt 2
zu Schreiben
II/43/1o

G e w i c h te

Bei R = 875o km Reichweite mit 3ooo kg Bomben d.i.
max.Reichweite von Focke Wulf ohne Außentank bei H_3.

	Me 264 (mit Normalprofil)		FW	Me 264/ 6m		
	Nach Baubeschreibung umgerechnet auf 4x 8ol E	Vergleichswerte n.FW-Rechnung ermittelt mit Festlast n.FW.	Nach Baubeschreibung	(mit Reichweit.Prof.) nach Baubeschr. umgerechnet auf 6 x 8ol E	(Mit Normalprofil) Vergl.Werte nach FW-Rechnung ermittelt m.Festlast nach FW.	(Mit Reichw.Profil) Vergl.Werte nach FW-Rechnung ermittelt mit Festlast nach FW.
Flugwerk	9758 kg	1o748 kg	15574 kg	1347o kg	13874 kg	13874 kg
Triebwerk ohne Kraftst.Behälter	6952	6952	8976	957o	957o	957o
Kraftst.Behälter	1735	1535	3264	224o	1946	178o
Ausrüstung	185o	1916	1916	252o	1916	1916
Bewaffnung u.Panzg.	3o1o(+GM 1)	422o	422o	41oo	422o	422o
Beatzg.u.Proviant	82o	7oo	7oo	9oo	7oo	7oo
Festlast	568o	6836	6836	7o2o	6836	6836
Landegw.o.Kraftstoffreserven	2414o	26o71	3465o	323oo	32226	32o6o
Kraftstoff und Schmierstoff	1591o	14o75	2285o	2o7oo	1782o	163oo
Bombenlast	3ooo	3ooo	3ooo	3ooo	3ooo	3ooo
Startgewicht	43o5o (> H_3)	43146 (> H_3)	605oo (H_3)	56ooo (> H_3)	53o46 (> H_3)	5136o (> H_3)

Augsburg,den 13.5.43
Hornung

Anlage I Blatt 3
zu Schreiben II/43/1o

	Me 264 (Normalprofil)		FW		Me 264/ 6m (Normalprofil)	
Durch Kraftstoff ausnützbares:	vorhanden	benötigt bei R= 875o km, i.Anlehnung an F.W.gerechnet	vorhanden	benötigt bei R =875o km	vorhanden	benötigt bei R =875o km in Anlehnung an F.W.gerechnet
Flügelvolumen	25,4 m^3	18,7 m^3	21,4 m^3	21,4 m^3	46,1 m^3	23,7 m^3
Rumpfvolumen o.Bombenraum	o,6 m^3		7,6 m^3	7,6 m^3	o,9 m^3	
Gesamtvolumen	26,o m^3	18,7 m^3	29,o m^3	29,o m^3	47,o m^3	23,7 m^3
vorhandenes Volumen f.Ausrüstg., Bewaffnung u.Besatzg.	ohne Heckstand 36 m^3	mit Heckstand 44 m^3	47 m^3		ohne Heckstand 45 m^3	mit Heckstand 46,5 m^3
Dem Beschuß von hinten ausgesetzte Kraftstoffbehälterfläche	i.Flügel: 14,3 m^2	1o,9 m^2	i.Flügel: 11,o m^2 i.Rumpf : 1,1 m^2		i.Flügel: 16,3 m^2	12,8 m^2

Augsburg,den 13,5.43

Hornung

Anlage I
Blatt 5
zu Schreiben
II/43/1o

		Me 264 (mit Normalprofil)		FW	Me 264/ 6m		
		Nach Baubeschreibung ungerechnet auf 4x8ol E	In Anlehnung an FW ermittelt mit Festlast n. FW.	Nach Baubeschreibung	(mit Reichw.Profil) nach Baubeschr. umgerechnet auf 6 x 8ol E	(mit Profil Me 264/ 4m)in Anlehnung an FW ermittelt mit Festlast n.FW.	(mit Reichweitenprof.) in Anlehnung an FW ermittelt mit Festlast n.FW.
Reichweitenflugzahl $\frac{n}{b} \cdot \frac{ca}{cw}$		7o,5	82,o	69,3	71,4	78,6	85,3
Reichweiten m.3 to Bo	R= konst.	G_{st}= 43o5o kg R= 875o km	G_{st}= 43146 kg R= 875o km	G_{st}= 6o5oo kg R= 875o km	G_{st}= 56ooo kg R= 875o km	G_{st}= 53o46 kg R= 875o km	G_{st}= 5136o kg R= 875o km
	bei H_3	G_{st}= 5oooo kg R= 1117o km	G_{st}= 5oooo kg R= 115oo km		G_{st}= 75ooo kg R= 1315o km	G_{st}= 75ooo kg R= 158oo km	G_{st}= 75ooo kg R= 158oo km
	m.Ueberlast bei H_2	G_{st}= 56ooo kg R= 1284o km	G_{st}= 56ooo kg R= 1353o km		G_{st}= 83ooo kg R= 1472o km	G_{st}= 83ooo kg R= 1659o km	G_{st}= 83ooo kg R= 176oo km
Reichweiten o.Bo als Fernstaufkl.	mit Ueberl. bei H_2	G_{st}= 56ooo kg R= 1413o km	G_{st}= 56ooo kg R= 149oo km	G_{st}= 6o5oo kg R= 1oooo km	G_{st}= 83ooo kg R= 1565o km	G_{st}= 83ooo.kg R= 1715o km	G_{st}= 83ooo kg R= 1859o km
		1ooo kg Festlaständerung bedeutet~6oo km Reichweitenänderung } bei jeweils gleichem Startgewicht					
		1ooo kg Abwurflaständerung " ~35o km "					
v_{max} i.VH 6,6 km mit N_{Ka} bei 1/2 B u.1/2 G_{Bo}		G_m = 3385o kg v = 595 km/h	G_m= 3485o kg v= 615 km/h	G_m = 47ooo kg v = 6o2 km/h	G_m = 445oo kg v = 612 km/h	G_m = 42946 kg v = 633 km/h	G_m = 4196o kg v = 638 km/h
Steiggeschw. i. Om m.N_{steig}	bei R= 875o km	w = 5,53 m/s.	w = 5,65 m/s.	w = 5,5 m/s.	w = 7,23 m/s.	w = 7,8 m/s.	w = 8,65 m/s.
	bei H_3	w = 3,86	w = 4,o7		w = 3,69	w = 3,51	w = 3,98
Startrollstrecke bei N_{start} mit Methanol	bei R=875okm	Sr= 125o m	Sr = 126o m	Sr =12oo m	Sr= 96o m	Sr= 85o m	Sr= 8oo m
	bei H_3	Sr= 2ooo m	Sr = 2ooo m		Sr= 22oo m	Sr= 22oo m	Sr= 22oo m
Landegeschwindigkeit mit Kraftstoff f.2 h Flug		G = 2648o kg v = 155 km/h	G = 28141 kg v = 159 km/h	G = 38ooo kg v = 16o km/h	G = 3535o kg v = 155 km/h	G = 34846 kg v = 154 km/h	G = 3446o kg v = 153 km/h

Augsburg,den 13.5.43

Hornung

Anlage I Blatt 4
zu Schreiben II/43/1o

I/4

Appendix 2

'Me 264 Fernaufklärer' Baubeschreibung – translated as 'The Me 264 Long-Range Reconnaissance Aircraft':
(RTP Translation No. GDC 15/654T – Ministry of Supply), 15 March 1943

RESTRICTED

R.T.P. TRANSLATION NO. G.D.C. 15/654T.
Issued by the Ministry of Supply.

The Me. 264 Long Range Reconnaissance Aircraft.

15th March, 1943.

Contents.

Foreword.

In its latest form the Me. 264 represents a 4 engined long range reconnaissance aircraft with a maximum range of about 15,000 km.

The construction of the airframe is essentially the same as for the types dealt with in the structural descriptions of April and December, 1941.

Part of the subsequent modifications were due to the new purpose of employment, part for the purpose of speeding up the series production of the aircraft.

In place of the special power units with DB 603 and Jumo 213 engines uniform units with BMW 801 engines are used. Apart from the air-cooling this engine type has the advantage of being more fully developed for Service employment compared with the engines provided earlier. The Me. 264 V-1 was fitted with Ju 88 power units with Ju 211J. Plans for pressurizing were abandoned.

The bomb space is used to accommodate radio-buoys, marker flares, dummy bombs, a GM-1 unit and three cine-cameras Rb 50/30. For the eventual carriage of bombs it will be possible to fit a rack for a total load up to 2,000 kg. as equipment.

The armament is hand-operated, and consists of A-gun position, B-1 and B-2 positions, C-position and two retractable lateral positions.

Elastic Leg.

Main undercarriage: VDM oleo-pneumatic 8 - 1130. 00.
Forward tricycle KC Oleo-pneumatic

Wheels

Main undercarriage 1550 x 575
Additional undercarriage 1550 x 575
(Intermediate solution)
Additional undercarriage 1550 x 575
(Final solution) 935 x 345
Front wheel 935 x 345

Tail Unit

Elevator unit with rudders attached as end plates. With this tail arrangement a good mark rearward field of fire is obtained for the B-2 gun position.

Tail plane adjustable in flight by means of an electrical adjustment system. Trimming flaps for elevator and rudder.

Control

Two rudder pedals for each of the two pilots' seats. Hand wheel swivelled over arbitrarily to one of the two seats.

Patin 3 - axis control (PDS).

/Wing

Wing Unit

Structure: single spar with forward and after auxiliary spar.
Subdivision: 2 inner wing sections, 2 outer sections and tip caps or wing extensions for 2 different wing spans. Junction positions with multiple tongue and groove fittings. Connection between inner sections and main spar with fork fitting.

Landing flap as split flap along the whole inner wing section. Ailerons on outer wing sections. No forward flap.

I. Technical Details

Design: Fuselage Structure

Circular cross-section, consists of 4 quarter-shells. Wing root forces transmitted through the main spar and two auxiliary bulkheads into the fuselage.

Sub-Division of Fuselage.

Front cockpit with all round view (operations room).
Bomb space with over-head passage including galley, rest-room and apparatus.

Rear cabin (arms and instruments room).

Accommodation in the fuselage only for the crew, jettisonable load, parts of the lubricant plant, GM-1 unit and cameras.

Undercarriage

Main undercarriage retractable. Forward tricycle unit retractable.

The main undercarriage travels sideways inwards into the wheel recess in the wing, which is closed by flaps lying flush with the skin. Forward undercarriage retracted with turn into the wheel recess below the pilots' cockpit, which is closed by flaps.

For taking-off with full load an auxiliary undercarriage for the main undercarriage is necessary. Provisional solution for 49 ton fully loaded weight:- an additional wheel fitted parallel to each fixed wheel: jettisoned with parachute. Final solution for 56 ton fully loaded weight:- tandem arrangement with two additional wheels to each fixed wheel (bicycle-wise), released at moment of becoming air-borne, and remains on the ground. The auxiliary undercarriage is braked and, when the aircraft has become airborne, rolls off the runway.

Power Unit

BMW 801 G-engine in the standard power unit. 801 TC unit contains: oil-radiator, hydr. radiator flap regulation, electro-mechanical and hydraulic airscrew operating system, control system, airscrew operating system without r.p.m. correction. Special r.p.m. ranges adjusted manually. Subsequently a higher power BMW 801E engine can be fitted in the 801 standard power unit. Wing frame designed to take this engine.

Fuel tanks in the wing, 2 additional jettisonable tanks outside the wing. Maximum content about 5000 kg. per tank. Tanks built in the wing, very easily accessible through screwed-on covers, they include self-sealing bag tanks holding 13830 kg. simple rubber bag tanks, holding 5850 kg,

Non-sealing tanks emptied by quick draining cocks in both wings. Ratio of the amount of the self-sealing to the non-sealing fuel is such, that with the maximum depths of penetration, sufficient fuel is available in the sealing tanks for return flight. Fuel scheme: empty first the outside tanks, then the non-sealing, lastly the self-sealing spare tanks in self-sealing service tanks. Oil in main tank in the fuselage, from which it is pumped into smaller service tanks for the individual engines.

GM-1 Plant

Supply for about 25 minutes flight. Tank housed in the bomb space. Injection through non-regulated 300 H.P. nozzles. Switch-on height 8,300 metres (attainable after about half the flight duration). Speed increase at about 8300 metres with a flying weight of 34,5 tons, about 95 km/hr. By fitting additional nozzles of smaller injection, it is possible to switch on at lower altitudes with larger flying weight.

Take-off Aids

R (Rocket) units to shorten the take-off run by increasing the starting thrust. 6 units, each of 1000 kg. thrust, under the wings, jettisoned after take-off, with parachutes. Mechanical release.

Hydraulic System

The following structural parts are operated electro-hydraulically:-

Undercarriage with covering
Landing flaps
Bomb Flap

Hydraulic switch easily accessible in the undercarriage recess.

Emergency operation by hand pump.

Safety and Life-saving Apparatus

High altitude breathing plant: 48 bottles each holding 2 lit O_2. Diaphragm lung, O_2 warning device, pressure gauge, stop valve and breathing tube combined on one board per seat. Gas mask holders by all seats. First-aid pack for crew of 8. One to two rubber dinghies in the fuselage tail. Release from rear emergency exit.

Signalling System.

Gun for firing signalling ammunition.

Electric Plant.

Aircraft network: 24 Volt
Aircraft battery for 45 amp. hr.
Aircraft network supply by generators, partly on the engine, partly mounted on instrument carriers operated by the engine unit.
Total output about 24 KW.
Main switch board behind the second pilot on the fuselage side wall.

Armament

Defensive armament:

A - position - 1 m.g. 131 in WL 131 (free mounting for A/C M/G 131).
B - 1 position - 1 m.g. 131 in DL 131.
B - 2 position - 1 m.g. 151 in HD 151 (hydraulic rotary ring mounting).
C - position - 1 m.g. 151 in special mounting.
D - positions - each 1 m.g. 151 in retractable mounting.
(side positions)

It is intended to replace the HD 151 in the B - 2 position by the HD 131 (indecipherable) with (indecipherable) MG 131.

\- 5 -

Bombs

In accordance with the purpose of employment, no special actual release system is provided. Subsequent installation of a rack for mixed load up to 2000 kg. in the bomb space is possible.

Armouring

Armouring for crew, gun positions and vital parts in the fuselage according to purpose of employment and main direction of attack behind and above to a weight of 1000 kg. Detailed arrangement not yet fixed. Weight of armour for the standard power unit included in the weight of the power unit.

Cameras

3X RB (cine) 50/30 in the bomb space. One mounted vertically, two obliquely on the sides, so that large width of exposure is obtained.

De-icing

Electric de-icing for wing and tail unit. Individual strips connected through period switch. Chemical de-icing for airscrews.

II. Structural details

Material

All materials used in the structure are A/C materials. Chiefly dural.

Strength.

H 4 with 38,9m span and 35 ton all up weight n = 4
H 3 with 43,0m span and 50 ton all up weight n = 2

Max. permissible landing weight: 28 ton.

With all up weight exceeding 50 t. the strength drops to the H.2 stress group, below which it does not fall, even with a max take-off weight of 56 ton.

Max. permissible speeds (near the ground).
V max. = 700 km/hr. (glide)
V h = 475 km/hr. (horizontal)
V h^{1} = 450 km/hr. (with bad visibility and near the ground)
V h^{11} = 300 km/hr. (with landing aids in operation)

Flight characteristics.

According to the flight characteristic directions.

Transport

The aircraft broken up can be transported on special lorries.

The outer wing sections only can be transported by rail.

Paint.

Foundation: aviation varnish 7122.
Seams puttied with putty DKM 100 10.
Wing coating and camouflage: aviation varnish 7109.

III. Weight analysis in kg. normal long range reconnaissance with bomb space.

1.	Wing unit	4972
2.	Fuselage	2773
3.	Tail unit	411
4.	Controls	143
5.	Undercarriage	1309

\- 6 -

6.	Power Unit	6200		
7.	Power unit installation	3342		
8.	Permanent equipment	1175		
9.	Additional equipment	516		
10.	Ballast	-		
11.	Paint	150		
12.				
13.	Reserves	159		
	As Civil aircraft, weight equipped.		21150	
14.	Armour (pilot's cockpit including reserve)		1000	
15.	Arms and ammunition		1200	
16.	Release gear		210	
17.	Photographic equipment		229 (interchange with 16 (? indecipherable)	
18.				
19.				
	As military aircraft, weight equipped		23360 ✓	
20.	Crew of 6	540 ✓		
21.	Provisions	100		
22.	Useful load (15 men with equipment)			
23.	Marginal fuel	200 ✓		
24.	Marginal oil	220 ✓		
25.	Coolant	-		
26.	G.M.1 filling	690 ✓		
	Landing weight		24410	
27.	Fuel in self-sealing tanks, without residue		✓ 13680	13680
28.	Fuel in non- " " " "		✓ 5800	5800
29.	Fuel in the bomb space		10350	3350
30.	Lubricant without residue		✓ 680	680
31.	Lubricant refill		✓ 580	580
32.	Bomb space tanks		✓ 500	500
33.	Bomb load		-	-
	Take-off weight		56000	48000
35.	Take-off aids		4120	4120
36.	Fuel for running and take-off		180	180

Initial running take-off weight with final additional undercarriage 60300
Initial running take-off weight with provisional additional undercarriage 53300

IV Dimensions — Engine data

Dimensions

Span	43,00 m
Length	20,90 m
Height	4,30 m
Wing area	127,70 m^2

Engine data

Gear ratio	1: 1,85
Take-off output. HP r.p.m.	1700/2700
(Climb and fighting output (HP/r.p.m. near ground	1490/2400
Climb and fighting output HP/r.p.m. at nominal altitude	1320/5300/2400

All outputs without dynamic pr. increase.

Airscrews.

VDM. 3 bladed (Ju 88) 3,70 diam.

/V.

\- 7 -

V. Performance

Preliminary comments

Weights. As a basis for the performance calculation a landing weight of 25,4 ton or 25t. is assumed for the aircraft fully fit for service. The possible starting weight from the point of view of performance is obtained from the performance charts according to the condition: 1m/sec. climb after taking off with maximum permissible sustained output.

Tolerances for the performance data.

With the fulfilment of the conditions specified below, the following tolerances with respect to the airframe are allowed:

for maximum speed: ± 3%
for rate of climb ± 4%
for range ± 10%

It should be noted that the ranges were calculated on the assumption of continuous throttling. Corrections were not introduced for throttling by stages.

Conditions applying to the engines.

BMW 801 G engine. Performance in accordance with diagram 9-801. 5072, dated 23.6.41.

Fuel consumption.

Figures where taken from the engine specifs. including tolerance for excess consumption (according to BMW. 9-801 5072 dated 23.6.41, permissible excess consumption + 2,5%). For long range flight at VH (full altitude?) accordingly a value was agreed for the mean consumption of 225 gr./H.P./hr, fuel + oil, + 2,5% tolerances on the assumption however that after reaching an altitude of 1 Km on boost, it was only after throttling down to the max, economic H.P. that climb to the service altitude was effected with this H.P. corresponding to the decrease in weight. This condition makes it possible to keep down the fuel consumption, as a long climb with maximum permissible sustained H.P. and the high consumption thereby entailed, is avoided. Possibility of unrestricted throttling during the long range flight correspondin[g] to the decrease in the H.P. required, is assumed. An output below the minimum permissible H.P. is not anticipated.

Engine performance

According to the accompanying performance curve.

Utilization of dynamic pressure.

75% assumed

Reaction

According to BMW 9-801, 5072 dated 23.6.41, e.g. 65,9 kg/1000 H.P. at ground level with take-off output.

Radiator drag

In the performance calculation, the drag of the radiator and mounting of the DB 603 H. (swivel radiator with 64m^2 frontal area) was retained.

According to the BMW data, the power absorption for the cooling is adequate[ly] covered by this assumption.

\- 8 -

Type

Range reconnaissance
Engines. 4X BMW 801 G.
Full armament
Armour 1000 kG
Bomb load: Nil.
3 X RB 50/30 Cine cameras.

	with final additional undercarriage	with present undercarriage
Take-off weight	56000 kg	49000 kg.
Initial take-off running weight	60300 kg	53300 kg.
Range with 25400 kg. Landing weight.	15000 km.	12500 km.
Length of take-off		
without additional thrust	-	2400 m
with 4000 kg additional thrust	2100 m	1500 m
Climb after take-off		
with boost	1,8m/sec.	3,2m/sec
with sustained output	1m/sec.	2,3m/sec

Max. Speed.
at 6,1 km altitude with boost.
with 36 ton. flying weight. 545 km/hr.
at 8,3 km with 34,5 ton. with nut G.M.1. 470km/hr.
at 8,3 with 24,5 ton. with G.M.1. 565 km/hr.

Service ceiling
with sustained H.P. with 36 ton. 8000 m

Endurance 45 hr.

Mean cruising speed 350 km/hr.

Landing speed (G=25,4t) 160 km/hr.
Wing loading: Landing (G=25,4t) 200 kg/m^2
Take-off (G=56t. 440 kg/m^2

Useful load.

When carrying 2000 kg. bomb load, the range is reduced by about 100 km.

Appendix to Description of the He. 264.

Long range reconnaissance aircraft as a special type for the experimenta[l] station for high altitude flight (V.f.H.)

The V.f.H. requires a number of long range reconnaissance aircraft with in addition to their actual purpose of employment are suitable for the transport a limited crew over long distances. The accommodation of 15 men on benches means relinquishing the bomb space, whereby the closed space necessary for th[e] transport of personnel is acquired.

The aircraft corresponds essentially to the normal long range reconnaiss[ance] machine already described, with the following differences.

1. The cameras, 3 X R.B. 50/30 are installed in the after cabin space.
2. The defensive armament is limited to the B-1 and B-2 gun positions.
3. The armour is omitted altogether.
4. The bomb space is omitted. Benches to seat 15 men are fitted instead.
5. Heating, heat insulation and de-icing systems are omitted.

\- 9 -

Weight analysis	Special V.f.H. reconnaissance without bomb space.	
1. Wing unit	4972	
2. Fuselage unit	2028	
3. Tail unit	411	
4. Control	143	
5. Undercarriage	1309	
6. Power unit	6200	
7. Permanent equipment	800	
8. Power unit installation	3342	
9. Additional equipment	516	
10. Ballast	-	
11. Paint	150	
12. -	-	
13. Reserves	629	
Weight equipped (civil)		20500
14. Armour		-
15. Arms & Ammunition		-
16. Bomb equipment		-
17. Cameras		250
18.		
19.		
Weight equipped - (military)		20750
20. Crew 8 men	720	
21. Provisions	100	
22. Useful load	-	
23. Marginal fuel	200	
24. Marginal Oil	220	
25. Coolant	-	
26. GM-1 filling	600	
Landing weight	25400	25400
27. Fuel in self-sealing tanks without marginal fuel	13680	13680
28. Fuel in non-self sealing tanks without margin	5800	5800
29. Fuel in droppable tanks	9360	9360
30. Oil without marginal oil	680	680
31. Oil for refill?	580	580
32. Droppable tanks	500	500
33. Jettisonable load	-	-
Flight take-off weight	56000	49000
34. Reserve		
35. Take-off aids	4120	
36. Fuel for run and take-off	180	
Initial running take-off weight with final additional undercarriage	60300	
" " " " " " temporary undercarriage		53300

/Performance

\- 10 -

Performance

Type Long range reconnaissance without bomb space for V.f.H.

Engines 4 x MBW 801 G.
Armament Nil (i)
Armour Nil
Bomb load Nil
Useful load 2000 kg.

Cameras 3 x RB 50/30 cine cameras.

(i) Installation of B-1 and B-2 gun positions reduces the range by 1000 km.

		with final add. undercarriage.	with temporary undercarriage.
Flight take-off weight		56000 kg.	49000 kg.
Initial running take-off weight		60300 kg.	53300 kg.
Range	With 25000 kg. landing weight	16500 km.(i)	14000 kg.(i)
Take-off length	without additional thrust		2400 m
	with 4000 kg additional thrust	2100	1500 m
Climb after take-off	with climb boost	1,8 m/sec	3,2 m/sec
	with sustained H.P.	1 m/sec	2,3 m/sec
Maximum speed	at 6,1 km. with climb boost, 36 ton flying weight		545km/hr
	at 8,3 km. with 34,5 ton without G.M.1.		470km/hr
	at 8,3 km. with 34.5 ton with G.M.1.		565km/hr
Service ceiling	with sustained H.P. with 36 ton weight		8000 m
Endurance			45 hr.
Mean cruising speed			350 km/hr
Landing speed	25 ton		160 km/hr
Wing loading	Landing 25 ton		196kg/m²
	Take-off 56 ton		440kg/m²

(i) If B-1 and B-2 gun blisters are mounted, the range is reduced by 1000 km.

Appendix 3

Status Sheet for Messerschmitt Me 264 V 1, W.Nr. 264 00001, RE+EN, 1 May 1944

Bauzustand v. 16.4.44 **Zustandsblatt Me 264 V** Werk Nr. 264 00001 **Blatt**

Tragwerk

Flügeldaten		Landehilfe		Querruder	Ruderausschläge	links	rechts	Flettnerausschl. autom.	links	rechts
F: 124,3 m²	V-Stellung: 4°	Art: 2 h Klappe	Startstellung 25°	F 2 x 2,7 m²	Querruder innen			Innenflettner		
b: 38,9 m		F: 2 x 8,6 m²		b 2 x 8 m	nach oben:	20°20	19°55	nach unten:	10°20	9°45'
Λ: 12,15		b: 2 x 10 m		L 471	nach unten	14°20	13°50	nach oben:	16°	14°50'
T: 0,238		%t ~ 22 %	Landestellung. 45°	La 204	Querruder außen			Außenflettner		
li: 4,90				Drehachse: 26 %	nach oben:	20°10	18°45'	nach unten:	11°	9°30'
					nach unten	14°30	15°20	nach oben:	15°	14°3[illegible]'

Höhenleitwerk

Daten		Ruderausschläge	Flettnerausschläge			Trimmung	Bemerkungen
F: 1[illegible],9 m²	Rudertiefe zuges. T 0,56	Drücken 30°15'	nach oben	autom.: 25°20'	hand: 9°20'	Flossenweg: + 4°	Flettneranschluß autom. 1:0,83
b: 7,28 m		Ziehen: 30°	nach unten	autom.: 25° bei Flossenstellung 0°	hand 7°35' - Ruderstellung 0°	- 6°	
Λ: 3,82							

Seitenleitwerk

Daten	Ruderausschläge	links	rechts	Flettnerausschläge autom.	links	rechts	hand	links	rechts	Bemerkungen
F. 2 x 4,35 m²	nach links	25°	25°20'	nach rechts	12°	12°	nach rechts	12°	12°20'	Flettneranschluß autom. 1:0,48
b: 7,28 m	nach rechts	25°	25°30'	nach links	9°	9°	nach links	11°20	11°30'	
Λ: 1,8										

Rumpf

Ausführung	Form	Bemerkungen
Kabinenart: Vollsichtkanzel		

Fahrwerk

Hauptfahrwerk		Bugrad	Bemerkungen
Federbein Typ: VDM W.Nr. lks. 103 rts. 104	Federweg:	Federbein Typ EC W.Nr. 93236	Bugradbremse eingebaut nach Änderungsanweisg. A 771, 790, 767
Federbeindruck: 37 atü		Federbeindruck: 19 atü	Notsporn Mtt.AG. A 1770 Z.
Reifengröße: 1550 x 575		Reifengröße: 935 x 345	
Reifendruck: 25 atü		Reifendruck: 5 atü	
		Federweg: 260 mm	

Triebwerk

Motor	Luftschraube	Kraftstoffbehälter	Schmierstoffbeh.	Bemerkungen
Typ: BMW 801 MG/2 TC-1 W.Nr. 31 3760 la	Typ: 3-fl. VDM Verstellschr. D = 3,4 m	Art: Sack	Art: Leichtmetall	
Typ: " " " " 31 3629 li	Baumuster: 9-12164	Zahl: 2 x 9	Zahl: 4	
Typ: " " " " 31 3742	Werk Nr: 06 263 la	Inhalt: 28000 l	Inhalt: 4 x 100	
Typ: " " " " 31 3746	Werk Nr: 06 228			
	Werk Nr: 06 221			
	Werk Nr: 06 247			

Anlage zum Erprobungsbericht Nr. 5 vom 1.5.44 — Sachbearb.: [signature] 931 — Datum:

Source Notes and Bibliography

Misc. Documents
Correspondence with Isolde Baur, March 2001.
Generalmajor C.A. von Gablenz, *Betrifft: Über der Arbeiten am Flugzeugmuster Me 261 und Me 264, Berlin, 18 Mai 1942*, via J.R.Smith.
AI2 (g) Report 2208, 12.12.43. (author collection).
Kriegstagebuch der Chef des Luftwaffenführungsstabes, July 1944, via David Irving.
Persönliches Kriegstagebuch des Generals der Flieger Werner Kreipe, 22.7-2.11.44, Air Historical Liaison Office, HQ, USAF, Washington DC.
Hans Bertram: *Projekt Ostasienflug, Berlin September 1944.*
Oberkommando der Luftwaffe, Chef der TLR Nr. 46442/44, 4.Dez. 1944: *Etatisierung des Sonderkommando Nebel.*
General der Aufklärungsflieger, Nr.338/45, 16.1.1945: *Abgrenzung der Aufgaben FAG 5 und Sonderkommando Nebel.*
Lw. Organisationsstab Genst.Gen. 2 Abt., *Notiz über Besprechung bei Chef Genst.Gen. 2 Abt am 12.2.45*, 12.2.45

National Archives, (formally PRO), London:
AIR40/203: Me 264 Aircraft.
AIR40/2168: AI2 (g) Reports 2375-3027: 'German Aircraft – New and Projected Types'.
AIR20/7709: AHB.6 Translation No. VII/140, Extracts from Conferences on Problems of Aircraft Production, August 1954 and AHB.6 Translation No. VII/137, Fighter Staff Conferences, 1944.
AIR27/2007 & 2008: Operations Record Books, 540 Squadron, RAF.
HW13/47: War Department Military Intelligence Service – German Technical Aid to Japan, 31 August 1942.
AIR20/7708: AHB 6 Translation No. VII/124 – Extract from report of the Goering Conference on Aircraft Production Programme, 23 May 1944.
AIR20/7711: GAF Policy during Second World War – A Review by Oberst Bernd von Brauchitsch, AHB.6 Translation VII/153, April 1956.

Messerschmitt A.G., Augsburg:
Protokoll Nr. 15, 9.4.42, Me 264 mit Ju 88 Triebwerken; USAAF, Me/264/Re/59 via IWM, London.
Vergleich der Leistungsrechnung der E-Stelle Rechlin und Fa. Messerschmitt A.G. v. Fernaufklärer Me 264 m. Jumo 211, 11.11.42, USAAF, Me/264/Re/74 via IWM, London.
Flugbericht Nr. 863/1, 4.1.43, *Flugbericht Nr. 873/2*, 29.1.43, *Flugbericht Nr. 875/3*, 3.2.43, *Flugbericht Nr. 894/4*, 28.2.43, *Flugbericht Nr. 900/5*, 16.3.43, .
Ausrüstungs- und Leistungszusammenstellung für Fernaufklärer Me 264 laut Rücksprache mit der Versuchstelle für Höhenflüge am 27.1.43., T-2 Hq AMC USAAF, R4088 F400-407, Me/264/Re/81 via IWM, London.
Projektübergabe Zusatzfahrwerk Me 264 mit Fallschirm, 9.3.43 (via Sengfelder).
Kontrollbericht Nr. 31/43, 15.3.43, via IWM, London.
Erprobungsbericht Nr. 2 vom 7.3.-23.3.43, Me/EB/264/Re/7E-243, via IWM, London.
Erprobungsbericht Nr. 5 vom 11.8.43 bis 16.4.44, via IWM, London.
Erprobungsbericht Nr. 6 17.4.-17.5.44 (author collection).
Me 264 - Inderungen aus der Flugerprobung und Weiterführung der Flugerprobung, 21.4.43, Me/264/Re/27, via IWM, London.
Projektübergabe Abwurfbehälter Me 264, XV/157, 30.8.43, via IWM, London.
Protokoll – Entwurf, XV/186, Me 264 Sonderaufklärer, Navigationsausrüstung (Einbau-Besprechung), 22.10.43, via IWM, London.
Projektübergabe Me 264 als Sonderfernaufklärer (VfH), XV/201, 12.11.43 (author collection).
Protokoll Nr. 23, 25.1.44: Me 264 – Bau von insgesamt 5 V-Flugzeugen, 25.1.44 (author collection).
Me 264 – Sonderfernaufklärer L-264 XV/207, 10.2.44, via IWM, London.
Me 264 – Gewichtsaufstellung: Sonderfernaufklärer XV/208, 10.2.44, via IWM, London.
Me 264 – Sonderfernaufklärer, XV/214, 28.4.44..
Projektübergabe Abwurfbehälter Me 264, XV/157, 30.8.43.
Kontrollbericht, 8.6.44

Deutsche Lufthansa A.G.:
Notiz über die Besprechung betreffend Me 264 und Me 261 in Augsburg am 25. April 1942 (Berlin, 29 April 1942); Air Docs. Div. T-2, R 2697, F 16

IWM London: miscellaneous document and microfilmed records:
'Einsatzaufgaben für Fernstflugzeuge', GL/A-Rü/Br. 208/42, 12 Mai 1942 (US Air Documents Division T-2 collection).
Combination of Reciprocated and Turbojet Engines in Me 264 Bomber (*Die Kombination von Otto-Motor und TL-Antrieb am Bespiel der Me 264*), Huber and von Stotzingen, BMW Flugmotorenbau, Munich, EZS Report No. 32, 29 October 1943 (HQ AMC, Wright-Patterson AFB, Translation No. F-TS-1566-RE, 16.9.1947).
Forschungsinstitut für Kraftfahrwesen und Fahrzeugmotoren and der Technisches Hochschule Stuttgart: Berichtsbrief 80 849/4 – Flatterversuche am Bugrad (Einzelrad) Me 264 im FKFS-Versuchsanhänger, 13.1.44, *Dr.-Ing.* E. Maier, Speer documents, Box 159/FD 4921/45.
The De-Icing of the Me 264: Messerschmitt Design Report, 23 September 1944 (RTP/TIB Translation, No. GDC. 15/256/T, issued by Ministry of Supply)

Articles and pamphlets etc
Messerschmitt-Me 264 – ein außergewöhnlicher Fernstaufklärer mit 15000 km Reichweite, Flugwelt, Flugwelt-Verlag GmbH, Wiesbaden, 1960
Garello, Giancarlo, *Obiettivo: New York – I progettati raid della Regio Aeronautica sugli Stati Uniti / Target New York – Italian planned raids to United States*, AeroFan, Anno 15 – N.62 – Lug.-Set.1997
Griehl, Manfred, *Das 'Bananaenflugzeug' – Die Entwicklungsgeschichte der Me 264 – Teil 1*, Flugzeug, 2/96
Griehl, Manfred, *Ein Traum zerbricht – Die Entwicklungsgeschichte der Me 264 – Teil 2*, Flugzeug, 3/96
Griehl, Manfred, *Die Me 264 V1 und ihre Enkel – Die Entwicklungsgeschichte der Me 264 – Teil 3*, Flugzeug, 4/96
Muscha, William R., *Strategic Elements in Interwar German Air Force Doctrine*, Army Command and General Staff College, Fort Leavenworth, Kansas, 2001
Overy, R.J., *From 'Uralbomber' to 'Amerikabomber': the Luftwaffe and Strategic Bombing*, The Journal of Strategic Studies, 1 (2), September 1978

Books
Baumbach, Werner, *Broken Swastika – The defeat of the Luftwaffe*, Robert Hale, London, 1986
Baur, Isolde, *A Pilot's Pilot – Karl Baur: Chief Test Pilot for Messerschmitt*, J.J. Fedorowicz Publishing, Inc., Winnipeg, 2000
Beauvais, Heinrich; Kössler, Karl; Mayer, Max & Regel, Christoph, *German Secret Flight Test Centres to 1945*, Midland Publishing, Hinckley, 2002
von Below, Nicolaus, *At Hitler's Side – The Memoirs of Hitler's Luftwaffe Adjutant 1937-1945*, Greenhill Books, London, 2001
Bekker, Cajus, *The Luftwaffe War Diaries*, Macdonald & Co., London, 1968
Bernstein, Jeremy, *Hitler's Uranium Club – The Secret Recordings at Farm Hall*, American Institute of Physics, New York, 1995 (Uncorrected Proof)
Blair, Clay, *Hitler's U-Boat War – The Hunted 1942-1945*, Weidenfeld & Nicolson, London, 1999
Boog, Horst, *Die deutsche Luftwaffenführung 1935-1945 – Führungsprobleme, Spitzengliederung, Generalstabsausbildung*, Deutsche Verlags-Anstalt, Stuttgart, 1982
Brooks, Geoffrey, *Hitler's Nuclear Weapons*, Leo Cooper, London, 1992
Brütting, Georg, *Das Buch der deutschen Fluggeschichte: Band 3*, Drei Brunnen Verlag, Stuttgart, 1979
Corum, James S., *The Luftwaffe – Creating the Operational Air War 1918-1940*, University Press of Kansas, Kansas, 1997
Duffy, James P., *Target: America – Hitler's Plan to Attack the United States*, Praeger Publishers, Westport, 2004
Ebert, Hans J., Kaiser, Johann B., Peters, Klaus: *Willy Messerschmitt: Pioneer of Aviation Design*, Schiffer Publishing, Atglen, 1999
Fleischer, Wolfgang, *German Air-Dropped Weapons to 1945*, Midland Publishing, Hinckley, 2004
Galland, Adolf, *The First and The Last*, Methuen & Co, London,1955
Gilbert, Martin, *Second World War*, Weidenfeld & Nicolson, 1989
Griehl, Manfred and Dressel, Joachim, *Zeppelin! – The German Airship Story*, Arms & Armour Press, London1990
Griehl, Manfred and Dressel, Joachim, *Heinkel He 177, 277, 274*, Airlife Publishing, Shrewsbury, 1998
Griehl, Manfred, *Luftwaffe over America – The Secret Plans to Bomb the United States in World War II*, Greenhill Books, 2005
Grunberger, Richard, *A Social History of the Third Reich*, Penguin Books, Harmondsworth, 1971
Heinkel, Ernst, *He 1000*, Hutchinson & co, London, 1956
Herwig, Dieter and Rode, Heinz, *Luftwaffe Secret Projects – Strategic Bombers 1935-1945*, Midland Counties Publications, Hinckley, 2000
Homze, Edward L., *Arming the Luftwaffe – The Reich Air Ministry and the German Aircraft Industry, 1919-39*, University of Nebraska Press, Lincoln and London, 1976
Irving, David, *The Virus House – Germany's Atomic Research and Allied Counter-measures*, William Kimber, London, 1967
Irving, David, *The Rise and Fall of the Luftwaffe – The Life of Erhard Milch*, Purnell Book Services, London 1973
Irving, David, *Hitler's War*, Hodder and Stoughton, London, 1977
Irving, David, *The War Path*, Michael Joseph, London, 1978
Van Ishoven, Armand, *Messerschmitt – Aircraft Designer*, Gentry Books, London, 1975
Kahn, David, *Hitler's Spies*, Hodder and Stoughton, London, 1978
Kay, Antony L. , *German Jet Engine and Gas Turbine Development 1930-1945*, Airlife Publishing, Shrewsbury, 2002
Kössler, Karl & Ott, Günther, *Die großen Dessauer – Junkers Ju 89, Ju 90, Ju 290, Ju 390 – Die Geschichte einer Flugzeugfamilie*, Aviatic Verlag, Planegg, 1993
Lee, Asher, *Goering – Air Leader*, Gerald Duckworth & Co., London, 1972
Ludwig, Paul, *P-51 Mustang: Development of the Long-Range Escort Fighter*, Classic Publications, Crowborough, 2003
Macksey, Kenneth, *Kesselring – The Making of the Luftwaffe*, David McKay Company, Inc., New York, 1978
McFarland, Stephen L. and Phillips Newton, Wesley, *To Command the Sky – The Battle for Air Superiority over Germany, 1942-1944*, Smithsonian Institution Press, Washington, 1991
Neitzel, Sönke: *Der Einsatz der deutschen Luftwaffe über dem Atlantik und der Nordsee 1939-1945*, Bernard & Graefe Verlag, Bonn, 1995
Nielsen, Gen.Lt. Andreas, *The German Air Force General Staff*, Arno Press, New York, 1959
Powers, Thomas, *Heisenberg's War: The Secret History of the German Bomb*, Jonathan Cape, London, 1993
Prien, Jochen, *IV./Jagdgeschwader 3: Chronik einer Jagdgruppe 1943-1945*, Eutin, 1996
Rust, Kenn C., *Fifteenth Air Force Story*, Historical Aviation Album, Temple City, 1976
Schramm, Percy E., *Kriegstagebuch des Oberkommandos der Wehrmacht 1940-1941, Teilband I*, Bernard & Graefe Verlag, München, 1982
Shirer, William L., *The Rise and Fall of the Third Reich*, BCA, 1978
Smith, J. Richard & Creek, Eddie J., *Me 262 Volume One*, Classic Publications, Burgess Hill, 1997
Smith, J. Richard, Creek, Eddie J., & Petrick, Peter, *On Special Missions – The Luftwaffe's Research and Experimental Squadrons 1923-1945*, Classic Publications, Hersham, 2003
Speer, Albert, *Inside the Third Reich*,Weidenfeld and Nicolson, London, 1970
Steel, Nigel and Hart, Peter, *Tumult in the Clouds – The British Experience of War in the Air, 1914-1918*, Hodder & Stoughton, London, 1997
Wohl, Robert, *A Passion for Wings – Aviation and the Western Imagination 1908-1918*, Yale University Press, New Haven and London, 1994

End Notes

Chapter One Megatheria

1. Wells, 279 and Wohl/75
2. Steel & Hart/284 and Griehl & Dressel/113
3. Ludwig/19
4. McFarland and Phillips Newton/28
5. Grunberger/16-17
6. Ebert, Kaiser, & Peters/92-93
7. Irving [Rise and Fall]/142
8. Irving [Rise and Fall]/142
9. van Ishoven/68
10. Homze/196
11. van Ishoven/67
12. Corum/175
13. Smith & Creek/21-26
14. Nielsen/155 and Overy/155
15. Corum/134-35
16. Macksey/44
17. Nielsen/156-57

Chapter Two 'The Banana Plane'

1. Von Below/20
2. Ebert, Kaiser & Peters/210
3. Ebert, Kaiser & Peters/216
4. Griehl – Luftwaffe over America (LOA)/21
5. Griehl – Me 264/1
6. Heinkel/215
7. Overy/157
8. Duffy/37 and www.mazal.org/archive/imt/03/IMT03-T389.htm
9. Bremer Nachrichten, 12 August 1938 (via Creek)
10. van Ishoven/115

Chapter Three 'Amerika Bomber'

1. Gilbert/100
2. Shirer/879
3. Brooks/37
4. Lee/97
5. Griehl/Me 264/1
6. Schramm/177 & Irving [Hitler's War]/186
7. Duffy/115-16 and www.mazal.org/archive/imt/03/IMT03-T389.htm
8. Griehl/Me 264/1
9. Letter Isolde Baur to author, 15 March 2001
10. Ebert, Kaiser & Peters/216, Griehl – Me 264/1 and Griehl – LOA/29
11. Griehl – LOA/30-31 and Herwig & Rode/102
12. Ebert, Kaiser & Peters/217
13. Griehl/Me 264/1
14. Shirer/875
15. Griehl – LOA/14
16. Griehl – LOA/33
17. Speer/121
18. Shirer/879
19. www.mazal.org/archive/ imt/03/IMT03-T390.htm
20. Overy/159-61
21. Irving [Hitler's War]/352
22. Grunberger/53
23. Irving [Hitler's War]/345
24. NA/AIR40/203
25. Griehl/Me 264/1 and Ebert, Kaiser & Peters
26. Griehl – LOA/65
27. Griehl/Me 264/1 and LOA/46
28. Garello/AeroFan No. 62/132-149
29. Messerschmitt AG, Augsburg, Protokoll Nr. 15, 9.4.42, Me 264 mit Ju 88 Triebwerken
30. Ishoven/163
31. Von Gablenz, Betrifft: Über der Arbeiten am Flugzeugmuster Me 261 und Me 264, 18 Mai 1942
32. Griehl/Me 264/3 and LOA/64
33. Smith & Creek/34
34. DLH A.G., Notiz über die Besprechung betreffend Me 264 und Me 261 in Augsburg am 25. April 1942 (29 April 1942)
35. Griehl/Me 264/1 and LOA/47
36. Duffy/49
37. Kdo der E-Stellen der Lw., Obstlt. und Kommandeur Petersen, re. Überprüfung der Arbeiten am Flugzeugmuster Me 264 an Verteiler, dated 7 May 1942, No. 15200/42 quoted in Ebert, Kaiser & Peters/217-18
38. Griehl/Me 264/1 and LOA/47
39. Beauvais, Kössler, Mayer & Regel/127
40. Kahn/157
41. 'Einsatzaufgaben für Fernstflugzeuge', GL/A-Rü/Br. 208/42, 12 Mai 1942
42. Duffy/38
43. Griehl/Me 264/1
44. Griehl/LOA/50-51
45. Baumbach/109
46. Irving [The Virus House]/213 and Powers/145 & 514
47. Forsyth/179 and Smith & Kay/564
48. Griehl/Me 264/1 and LOA/56-57
49. Duffy/51, Griehl/Me 264/1 and LOA/57-58
50. Griehl/LOA/58
51. Griehl/Me 264/1
52. Griehl/Me 264/1 and Ebert, Kaiser & Peters/220
53. Griehl/LOA/60
54. Griehl/Me 264/1
55. Griehl/Me 264/3 and Irving [Rise & Fall]/170
56. Overy/161
57. Neitzel/149
58. Griehl/Me 264/1
59. Griehl/Me 264/3

60. Griehl/Me 264/3
61. Mtt A.G., Vergleich der Leistungsrechnung der E-Stelle Rechlin und Fa. Messerschmitt A.G. v. Fernaufklärer Me 264 m. Jumo 211, 11.11.42

Chapter Four 'Sudenten'

1. Messerschmitt A.G., Augsburg: Flugbericht Nr. 863/1, 4.1.43
2. Baur/116
3. Griehl/Me 264/3
4. Van Ishoven/172
5. Griehl/Me 264/2
6. Garello/AeroFan No. 62/132-149
7. Flugbericht Nr. 873/2
8. Baur/116
9. Messerschmitt A.G. Augbsurg: Ausrüstungs- und Leistungszusammenstellung für Fernaufklärer Me 264 laut Rücksprache mit der Versuchstelle für Höhenflüge am 27.1.43.
10. Messerschmitt A.G. Augsburg: Flugbericht Nr. 875/3, 3.2.43
11. Griehl/Me 264/3
12. Messerschmitt A.G. Augsburg: Flugbericht Nr. 894/4, 28.2.43
13. Messerschmitt A.G., Augsburg, Flugbericht Nr. 900/5, 16.3.43
14. Griehl/Me 264/3
15. Messerschmitt A.G. Augsburg: Kontrollbericht Nr. 31/43, 15.3.43
16. Griehl/Me 264/3
17. Messerschmitt A.G. Augsburg: Kontrollbericht Nr. 31/43, 15.3.43
18. Griehl/Me 264/3
19. Griehl/LOA/89
20. Griehl/LOA/116 and Neitzel/161
21. Griehl/LOA/92
22. Griehl/Me 264/1
23. NA/AIR27/2008
24. NA/AIR40/203
25. NA/AIR20/7709
26. Irving/Rise & Fall/65
27. Baumbach/107
28. Baur/116
29. Griehl/Me 264/1&3
30. Messerschmitt A.G., Augbsurg: Me 264 - Inderungen aus der Flugerprobung und Weiterführung der Flugerprobung, 21.4.43
31. Griehl/LOA/96-114
32. Griehl/Me 264/2
33. Griehl/LOA/117, Messerschmitt A.G., Augsburg: Fortsetzung zum Schreiben an Herrn F.W. Seiler, Gkos II/43/10, (und Anlagen), 13&15.5.43 and Messerschmitt A.G., Augsburg: Leistungsvergleich der 4- und 6-motorigen Me- Fernkampfflugzeuge mit Fernkampfflugzeuge von Focke-Wulf, 15.5.43.
34. Galland/332
35. Griehl//Me 264/3 & LOA/94 & 117
36. Blair/311
37. Baumbach/107
38. Griehl/Me 264/2
39. Griehl/LOA/125-26
40. Griehl/Me 264/3
41. NA/AIR40/2168 AI2 (g) Reports 2375-3027: 'German Aircraft – New and Projected Types' and Messerschmitt-Me 264 – ein außergewöhnlicher Fernstaufklärer mit 15000 km Reichweite, Flugwelt, Wiesbaden, 1960, pg 1
42. Combination of Reciprocated and Turbojet Engines in Me 264 Bomber (Die Kombination von Otto-Motor und TL-Antrieb am Bespiel der Me 264), Huver and von Stotzingen, BMW Flugmotorenbau, Munich, EZS Report No. 32, 29 October 1943
43. Kay/220-24
44. Messerschmitt A.G., Augsburg , Erprobungsbericht Nr. 5 vom 11.8.43 bis 16.4.44

Chapter Five 'With just one or two bombs...'

1. Griehl/Me 264/2
2. Messerschmitt A.G., Augsburg: Projektübergabe Abwurfbehälter Me 264, XV/157, 30.8.43
3. Griehl/Me 264/2
4. Griehl/LOA/131 and 264/2
5. Messerschmitt A.G., Augsburg, Protokoll – Entwurf, XV/186, Me 264 Sonderaufklärer, Navigationsausrüstung (Einbau-Besprechung), 22.10.43
6. Messerschmitt A.G., Augsburg: Projektübergabe Me 264 als Sonderfernaufklärer (VfH), XV/201, 12.11.43
7. Dr.-Ing. E. Maier, Forschungsinstitut für Kraftfahrwesen und Fahrzeugmotoren and der Technisches Hochschule Stuttgart: Berichtsbrief 80 849/4 – Flatterversuche am Bugrad (Einzelrad) Me 264 im FKFS-Versuchsanhänger, dated 13.1.44,
8. NA/AIR40/203
9. Correspondence between J.R.Smith and Erich Sommer, 1997
10. Messerschmitt A.G., Augsburg, Protokoll Nr. 23, 25.1.44: Me 264 – Bau von insgesamt 5 V-Flugzeugen
11. NA/AIR40/203 – Me 264 Aircraft
12. Messerschmitt A.G., Augsburg: Me 264 – Sonderfernaufklärer L-264 XV/207, 10.2.44
13. Messerschmitt A.G., Augsburg: Me 264 – Gewichtsaufstellung: Sonderfernaufklärer XV/208, 10.2.44
14. Messerschmitt A.G., Augsburg, Erprobungsbericht Nr. 5 vom 11.8.43 bis 16.4.44
15. AI2 (g) Report 2208, 12.12.43.
16. NA/AIR27/2007: Ops Record Book, 540 Squadron
17. NA/AIR40/203: RAF Medmenham Interpretation Report No. L.167, 10.4.44
18. Messerschmitt A.G., Augsburg, Erprobungsbericht Nr. 5 vom 11.8.43 bis 16.4.44
19. Baur/116
20. Smith, Creek & Petrick/On Special Missions/13
21. Messerschmitt A.G., Augsburg, Abschrift/FAM/Ka. – Aktenvermerk vom 16.4.44., Me 264 Nr.101: Betrifft: Nachfliegen Me 264 durch H. Oberstltn. Kneemeier [sic] ind Flugbaumeister Scheibe in Oberammergau am 16.4.44.
22. Messerschmitt A.G., Augsburg, Erprobungsbericht Nr. 5 vom 11.8.43 bis 16.4.44
23. Messerschmitt A.G., Augsburg, Abschrift/FAM/Ka. – Aktenvermerk vom 16.4.44., Me 264 Nr.101: Betrifft: Nachfliegen Me 264 durch H. Oberstltn. Kneemeier [sic] ind Flugbaumeister Scheibe in Oberammergau am 16.4.44.
24. Messerschmitt A.G., Augsburg: Erprobungsbericht Nr. 6 17.4.-17.5.44
25. Ebert, Kaiser & Peters/222
26. Griehl/Me 264/3 and Erprobungsbericht Nr. 6 17.4.-17.5.44
27. Griehl/Me 264/3
28. Messerschmitt A.G., Augsburg, 'Me 264 – Sonderfernaufklärer', XV/214, 28.4.44.
29. NA/HW13/47 War Department Military Intelligence Service – German Technical Aid to Japan, 31 August 1942, pg 134
30. NA/AIR20/7708, AHB 6 Translation No. VII/124 – Extract from report of the Goering Conference on Aircraft Production Programme, 23 May 1944
31. Irving [Rise & Fall]/284
32. Messerschmitt A.G. Erprobungsbericht Nr. 7 18.5-28.6.44
33. Messerschmitt A.G. Kontrollbericht, 8.6.44
34. Griehl/Me 264/3
35. Griehl/Me 264/2
36. Baumbach/110
37. NA/AIR20/7709, AHB.6 Translation No. VII/137, Fighter Staff Conferences, 1944
38. Messerschmitt A.G. Erprobungsbericht Nr. 7 18.5-28.6.44
39. Griehl/Me 264/2
40. Der Chef des Luftwaffenführungsstabes, Ia Nr. 4532/44, 9.7.44
41. Griehl/LOA/195
42. Prien/189 and Rust/31
43. Messerschmitt A.G. Erprobungsbericht Nr. 7 18.5-28.6.44
44. NA/AIR40/2168
45. Ministry of Supply translation: The De-Icing of the Me 264, 23.9.44
46. Griehl/Me 264/2
47. Kreipe KTB, 21.8.44
48. Kreipe KTB, 21.8.44
49. Griehl/Me 264/2
50. Bertram: Projekt Ostasienflug September 1944
51. NA/HW13/47 War Department Military Intelligence Service – German Technical Aid to Japan, 31 August 1942, pg 134-45
52. Neitzel/229 and Griehl/Me 264/2
53. Griehl/Me 264/2 and LOA/199-200
54. OKL, Chef der TLR Nr. 46442/44, 4.Dez. 1944
55. Gen der Aufklärungsflieger, Nr.338/45, 16.1.1945
56. Lw. Organisationsstab Genst.Gen. 2 Abt., 12.2.45
57. Overy/163
58. NA/AIR20/7711

Chapter Six 'Myth and Reality'

1. NA/AIR40/203
2. NA/AIR40/203
3. Irving [The Virus House]/107-09 & 279
4. Speer/225
5. Bernstein/xxiii
6. Bernstein/xxiii
7. Irving [The Virus House]/94
8. Bernstein/xxiv
9. Powers/153
10. Powers/147
11. Speer/226
12. Powers/517
13. Irving [The Virus House]/109
14. Irving [The Virus House]/109
15. Irving [The Virus House]/267
16. Brooks/38
17. Irving [The Virus House]/268
18. NA/AIR40/203
19. Overy/155
20. Overy/165
21. Bekker/374-75
22. Nielsen/152

Index

(Wherever possible, last known ranks and positions are given)